T0074220

Lecture Notes in Social Networks

Series editors

Reda Alhajj, University of Calgary, Calgary, AB, Canada
Uwe Glässer, Simon Fraser University, Burnaby, BC, Canada
Huan Liu, Arizona State University, Tempe, AZ, USA
Rafael Wittek, University of Groningen, Groningen, The Netherlands
Daniel Zeng, University of Arizona, Tucson, AZ, USA

Advisory Board

Charu C. Aggarwal, Yorktown Heights, NY, USA
Patricia L. Brantingham, Simon Fraser University, Burnaby, BC, Canada
Thilo Gross, University of Bristol, Bristol, UK
Jiawei Han, University of Illinois at Urbana-Champaign, Urbana, IL, USA
Raúl Manásevich, University of Chile, Santiago, Chile
Anthony J. Masys, University of Leicester, Ottawa, ON, Canada
Carlo Morselli, School of Criminology, Montreal, QC, Canada

More information about this series at http://www.springer.com/series/8768

Reda Alhajj • H. Ulrich Hoppe • Tobias Hecking
Piotr Bródka • Przemyslaw Kazienko
Editors

Network Intelligence Meets User Centered Social Media Networks

 Springer

Editors
Reda Alhajj
Department of Computer Science
University of Calgary
Calgary, Alberta, Canada

Tobias Hecking
Computer Science and Applied
Cognitive Science
University of Duisburg-Essen
Duisburg, Nordrhein-Westfalen, Germany

Przemyslaw Kazienko
Faculty of Computer Science
and Management
The Engine Centre
Wroclaw University of Science
and Technology
Wroclaw, Poland

H. Ulrich Hoppe
Department of Computer Science and
Applied Cognitive Science
University of Duisburg-Essen
Duisburg, Nordrhein-Westfalen, Germany

Piotr Bródka
Department of Computational Intelligence
The Engine Centre, Faculty of
Computer Science and Management
Wroclaw University of Science
and Technology
Wroclaw, Poland

ISSN 2190-5428   ISSN 2190-5436   (electronic)
Lecture Notes in Social Networks
ISBN 978-3-319-90311-8   ISBN 978-3-319-90312-5   (eBook)
https://doi.org/10.1007/978-3-319-90312-5

Library of Congress Control Number: 2018944599

© Springer International Publishing AG, part of Springer Nature 2018
This work is subject to copyright. All rights are reserved by the Publisher, whether the whole or part of
the material is concerned, specifically the rights of translation, reprinting, reuse of illustrations, recitation,
broadcasting, reproduction on microfilms or in any other physical way, and transmission or information
storage and retrieval, electronic adaptation, computer software, or by similar or dissimilar methodology
now known or hereafter developed.
The use of general descriptive names, registered names, trademarks, service marks, etc. in this publication
does not imply, even in the absence of a specific statement, that such names are exempt from the relevant
protective laws and regulations and therefore free for general use.
The publisher, the authors and the editors are safe to assume that the advice and information in this book
are believed to be true and accurate at the date of publication. Neither the publisher nor the authors or
the editors give a warranty, express or implied, with respect to the material contained herein or for any
errors or omissions that may have been made. The publisher remains neutral with regard to jurisdictional
claims in published maps and institutional affiliations.

This Springer imprint is published by the registered company Springer Nature Switzerland AG
The registered company address is: Gewerbestrasse 11, 6330 Cham, Switzerland

Contents

Part I
Centrality and Influence

Targeting Influential Nodes for Recovery in Bootstrap Percolation on Hyperbolic Networks

Christine Marshall, Colm O'Riordan, and James Cruickshank

Abstract The influence of our peers is a powerful reinforcement for our social behaviour, evidenced in voter behaviour and trend adoption. Bootstrap percolation is a simple method for modelling this process. In this work we look at bootstrap percolation on hyperbolic random geometric graphs, which have been used to model the Internet graph, and introduce a form of bootstrap percolation with recovery, showing that random targeting of nodes for recovery will delay adoption, but this effect is enhanced when nodes of high degree are selectively targeted.

Keywords Bootstrap percolation · Bootstrap percolation with recovery · Hyperbolic random geometric graphs

1 Introduction

In this work, we examine agent-based modelling of Bootstrap Percolation on Hyperbolic Random Geometric graphs, and introduce a modified version of Bootstrap Percolation with Recovery on the same set of graphs. Bootstrap percolation has been used to model social reinforcement, with the rationale that individuals are more likely to adopt a new technology, for example, if a significant number of their contacts already use that technology. A key question in bootstrap percolation is whether the activity will percolate or not. Our focus is on the effects of spatial structure on the spread or percolation of an activity, specifically observing bootstrap percolation on hyperbolic random geometric graphs. Our motivation for using hyperbolic networks is they share many features with real-world complex

C. Marshall (✉) · C. O'Riordan
Discipline of Information Technology, National University of Ireland, Galway, Ireland
e-mail: c.marshall1@nuigalway.ie; colm.oriordan@nuigalway.ie

J. Cruickshank
School of Mathematics, National University of Ireland, Galway, Ireland
e-mail: james.cruickshank@nuigalway.ie

© Springer International Publishing AG, part of Springer Nature 2018
R. Alhajj et al. (eds.), *Network Intelligence Meets User Centered Social Media Networks*, Lecture Notes in Social Networks,
https://doi.org/10.1007/978-3-319-90312-5_1

networks; typically displaying power-law degree distribution, high clustering, short path lengths, high closeness and significant betweenness centralisation.

We have previously created a set of hyperbolic random geometric graphs from unconnected, with increasing edge density, to fully connected. In this work, we attach a population of agents to the graphs and explore the effect of percolation processes on these agents on this set of graphs. Our first question is whether we can identify a distinct percolation threshold as we increase the number of edges, above which the activity percolates and below which the activity fails to percolate.

Typically work on bootstrap percolation has looked at the effect of the initial active seed set on the final outcome of the process, looking at the choice of active seed, or at the fraction of active seeds required to guarantee percolation of the activity. This has been used in, for example, viral marketing to see which nodes might best be targeted to optimise the spread of information. Our overall interest is in targeting nodes which might prevent the percolation of the activity, given a random seed set of fixed size. This could model a situation where a network is randomly targeted with active seeds, and we wish to know which nodes might have the best chance of obstructing the process. Our focus is on minimising the spread of an activity given a small-scale attack at random points within the network.

We develop a modified version of standard bootstrap percolation which allows for the recovery of a defined percentage of active nodes after each activation step; by recovery we mean the transition from active state back to inactive state. Our next questions are whether this will impact the spread of the activity, and, furthermore, if we selectively target active nodes of high degree, will this have a greater impact on the spread of activity than the same percentage chosen at random, given that hub nodes are commonly regarded as influential nodes in a network.

Section 2 provides related information about bootstrap percolation and hyperbolic graphs. Section 3 introduces our conceptual framework for Bootstrap Percolation with recovery. Section 4 describes our experimental set-up, and our results are presented in Sect. 5. Our conclusions are discussed in Sect. 6, together with suggestions for future work.

2   Background

In this section we describe bootstrap percolation and hyperbolic random geometric graphs.

2.1   Bootstrap Percolation

Bootstrap percolation is a dynamic process in which an activity can spread to a node in a network if the number of active neighbours of that node is greater than a predefined activation threshold. This process can model forms of social reinforcement where the spread of an activity largely depends on a tipping point in the number of activated contacts. These activities can include the spread of opinions or the adoption of a new technology; the underlying assumption is that people are more likely to adopt a new activity if several of their contacts are already involved in the same activity. The idea of Bootstrap Percolation was introduced by Chalupa, Leath and Reich in their 1979 work on Bethe lattices studying the mechanisms of ferromagnetism, or how materials become magnetised [13].

The bootstrap percolation model is essentially a two-state model, with agents either active or inactive. In some of the literature, these states are termed infected or uninfected, but basically means that an activity is present or not present. Standard bootstrap percolation is a discrete-time process, where the activation mechanism occurs synchronously for all nodes in the graph in rounds, at each time step. Starting with an inactive population of agents, a set of nodes is chosen as the active seed set, this choice may be at random or deterministic. An integer value activation threshold is chosen and the activation mechanism occurs at each time step, whereby a node becomes activated if the number of its active neighbours is at least that value. The rounds of activation are then repeated until equilibrium, where no further state change is possible, or at some predetermined parameter, such as the number of rounds. In the spatial form of the process, agents are attached to nodes in a network and activation depends on the number of active nodes directly connected to a node.

Bootstrap percolation has been studied on a variety of random graphs and complex networks [2–4, 6, 19, 34]. A key question is whether the activity will completely percolate. A lot of research has concentrated on the relationship between the initial active seed set and the final outcome of the process; this has in essence involved looking at the choice of the initial active seed set to optimise the spread of the activity [20], or the size of the initial seed set to ensure percolation [11, 17]. Kempe et al. investigated algorithms for optimising this set selection on a variety of networks, noting that this was NP-hard [21]. The bootstrap percolation process has been used to model information diffusion [23–25] and viral marketing [15], behavioural diffusion [12], such as opinion formation and voter trends, the adoption of brands, technology and innovation in social networks [16, 18, 31], and also cascading failures in power systems.

In the standard form of bootstrap percolation, activated nodes must remain active, no reverse state is allowed. In 2014, Coker and Gunderson investigated the idea of bootstrap percolation with recovery on lattice grids based on an update rule that infected nodes with few infected neighbours will become uninfected [14]. They used a probabilistic approach to determine thresholds for the probability of percolation based on the size of the initial active seed set. By examining various configurations of 2-tiles (pairs of sites that share an edge or a corner in the lattice grid), they determine the critical probabilities for percolation.

In epidemiology, related models are the SI and SIS models of disease propagation [32] which have similar state changes to bootstrap percolation and bootstrap percolation with recovery, respectively. In the SI model, the population is compartmentalised into either Susceptible (i.e. uninfected) or Infected sinks. Each susceptible individual has a probability of becoming infected, based on the transmission rate, which is a function of the typical number of contacts per individual, and of the infectivity of the disease. Once infected, individuals move to the infected compartment and remain infected. The SIS model is an extension of this model where infected individuals will recover (and become susceptible again) based on the rate of recovery.

The key point of interest is the epidemic threshold, and the effects of varying rates of infection and recovery on the spread of the disease. In the SI model, the infection will ultimately spread to the entire population. In the SIS model, for high recovery rates the illness will die out in the population and for low recovery rates the illness will become endemic [5]. These models are compartmental models, where the population in each sink is homogeneous; each individual within a compartment has the same likelihood of changing state.

Bootstrap percolation is distinctly different from the epidemic models, as the population of agents is heterogeneous, each with local knowledge of node properties in their neighbourhood. Once the initial active set has been chosen, the spread of the activity is determined by the number of neighbourhood links to active nodes, which is a function of the underlying network topology.

To address issues of network structure in the epidemic models, Pastor-Satorras and Vespignani in 2001 [27] introduced new compartments based on node degree, with nodes of the same degree having the same likelihood of changing state. In large-scale free models, the hub nodes ensure that the infection is likely to spread [5]. However, this approach does not take into account community structure or local clustering. Agent-based modelling of bootstrap percolation on a network is a simple dynamic process ideally suited to accounting for individual node features.

2.2 Hyperbolic Random Geometric Graphs

In Random Geometric Graphs, pairs of nodes are connected if they lie within a specified distance parameter of each other. They were developed in the 1970s to model situations where the distance between nodes is of interest, as in modelling the spread of a forest fire. These graphs have generally been created in the Euclidean plane, particularly the unit square and unit disc. In computer science, these models are frequently used to model wireless ad-hoc and sensor networks [7], where proximity between nodes defines network connections. Other uses include modelling neural networks [9], mapping protein-protein interactions [28], and in percolation theory [11], modelling processes such as diffusion in a network, fluid percolation in porous materials, fracture patterns in earthquakes and conductivity [30].

Hyperbolic random geometric graphs were developed by Krioukov et al. in 2010 [22]. In this model, pairs of nodes are connected if the hyperbolic distance between them is less than a specified distance parameter. To highlight the contrast between hyperbolic and Euclidean random geometric graphs, the Poincaré disc model transforms the negative curvature of the hyperbolic plane to a 2 dimensional disc. Points within the disc are not uniformly distributed; the hyperbolic model has more

Fig. 1 Claudio Rocchini
*Order-3 heptakis heptagonal
tiling* 2007, distributed under
a CC BY 2.5 licence [29]

"room" than the Euclidean disc. As the radius of a Euclidean disc increases, the circumference increases linearly; in a hyperbolic disc, the circumference increases exponentially. For example, a hyperbolic tree can represent more data as child nodes have as much space as parent nodes. This tends to give a fish-eye view to nodes within the disc, with greater emphasis placed on central nodes, while outer nodes are exponentially distant, this effect is captured in an image by Claudio Rocchini [29], see Fig. 1.

Hyperbolic geometric graphs have been used to explore structural properties and processes in complex networks, such as clustering [10] and navigability. Krioukov et al. have proposed that these models have potential for modelling the Internet graph, which they suggest has an underlying hyperbolic geometry [22]. Recent work by Papadopoulos et al. has developed a tool to map real-world complex networks, such as the Autonomous Systems Internet, to a hyperbolic space using the hyperbolic geometric model [26]. Recently, researchers have developed faster algorithms for creating large hyperbolic networks. Von Looz et al. have developed a speedy generator to create representative subsets of hyperbolic graphs, allowing for networks with billions of edges, while retaining key features of hyperbolic graphs [33], while Bringmann et al. have developed a generator to create a generalised hyperbolic model in linear time, by avoiding the use of hyperbolic cosines [8].

3 Conceptual Framework for Bootstrap Percolation with Recovery

The concept of bootstrap percolation with recovery allows for an activated node to become deactivated. This might model a change of mind situation, in trend adoption or behavioural diffusion, where individuals are initially interested in an activity popular with their contacts, but immediately change their mind and no longer endorse the activity, at least in the short term. In standard bootstrap percolation, a lot of work focuses on optimising the size and selection of the initial active seed set to ensure percolation. Our own focus is on a fixed-size randomly chosen seed set, motivated by the notion of a small-scale random seeding in a network, with a view to targeting recovery to inhibit the spread of the activity. Our proposed method of recovery involves targeting active nodes with specific node properties, such as high degree centrality; our aim is to identify those properties that might have the most impact on delaying the percolation process. We have chosen node degree, in the first instance, as node degree is a standard measure of influence in networks, [1, 5], and the hyperbolic geometric graphs display highly skewed degree centralisation, reflecting the high level of variance in node degree within each graph. We examine the effect that this has on the percolation of the activity, in comparison with the standard bootstrap percolation process, in which recovery is not permitted. As a control, we also simulate bootstrap percolation with random selection of active nodes for recovery.

3.1 Proposed Method of Bootstrap Percolation with Recovery

Our proposed method involves setting up the experiments as for the standard bootstrap process but introducing a recovery mechanism to follow the activation mechanism in each time step.

1. Set Parameters

    - Set size of active seed set A_o
    - Set integer value activation threshold r
    - Set recovery rate percentage $RR\%$

2. Define Mechanisms

    - Activation mechanism

      • For all inactive nodes, activate if the number of active neighbours is at least r.

    - Recovery mechanism

      • For all active nodes, select $RR\%$ to deactivate, by random or deterministic selection

3. Define Process

    - For each graph, attach an agent to each node in the graph
    - Set all agents inactive
    - Select active seed set
    - Apply activation mechanism followed by recovery mechanism at each time step until equilibrium

4 Experimental Set-up

Our simulations involve creating a set of hyperbolic random geometric graphs and observing the effect of graph topology on the spread of activity by embedding the bootstrap percolation process on these graphs.

A set of hyperbolic geometric graphs, each with 1000 nodes, was created with varying edge density from disconnected to fully connected, with 20 graphs created at each value of the distance parameter. Bootstrap percolation was embedded on this set of graphs and then our modified versions of bootstrap percolation with recovery, both random and targeted, were simulated on the same set of graphs to compare outcomes. The recovery experiments were repeated with varying recovery rate percentages ($RR\%$) of 10–90%, in steps of 10%.

4.1 Graph Creation

In this research, hyperbolic random geometric graphs are created using the model outlined by Krioukov et al. [22]. This approach situates the graph in a disc of radius R within the Poincaré disc model of hyperbolic space. Nodes in the graph have polar coordinates (r, θ) and are positioned by generating these coordinates as random variates; edges are created between pairs of nodes where the hyperbolic distance between them is less than the radius R of the disc of interest, varying R from 0.1 to 12 in increments of 0.1, to create graphs with an increasing number of edges, from disconnected to fully connected, with 20 graphs created at each distance parameter R. The relationship with R and edge density is not linear, see Fig. 2; therefore, R is used in all our results for ease of comparison.

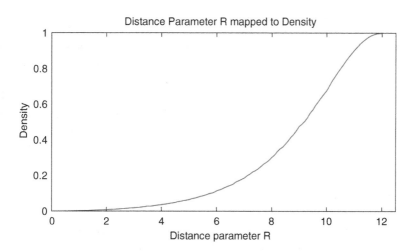

Fig. 2 Density parameter R mapped to edge density

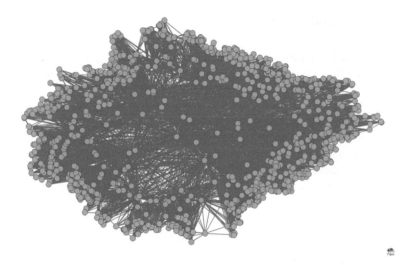

Fig. 3 Hyperbolic random geometric graph with 1000 nodes and an edge density of 0.037

Figure 3 shows a hyperbolic random geometric graph with 1000 nodes, distance parameter $R = 4$, and an edge density of 0.037. This shows a tree-like structure, with the major hub nodes in the centre and the majority of nodes, that is, those of lower degree, are situated towards the boundary.

4.2 Simulation of Bootstrap Percolation on the Set of Hyperbolic Graphs

A population of agents, engaged in the process of bootstrap percolation, is attached to each graph to observe any spread in activity. An agent is assigned to each node, with all agents initially inactive and, from these, 20 nodes are randomly selected to form the initial active seed set, A_o. The activation mechanism involves an activation threshold r whereby an inactive node becomes active if it has at least r active neighbours. At each time step, this activation mechanism is applied synchronously to all nodes in the graph and is repeated until no further state change is possible and a stable equilibrium is reached; the number of active nodes is recorded at this point. Our experiments are then repeated varying the activation threshold r from 2 to 10.

4.3 Bootstrap Percolation with Recovery

The bootstrap percolation process is modified to allow for reverse state change from active to inactive, as outlined in Sect. 3.1 by introducing a recovery phase in each time step, immediately following activation. Our first set of experiments involves random selection of active nodes for recovery. The experiments are then repeated targeting active nodes of high degree for recovery. In both sets of simulations, the experiments are repeated with varying Recovery Rate percentages ($RR\%$) from 10 to 90% in steps of 10% for each activation threshold in the range 2–10, as before.

As for the standard process, a population of agents, engaged in the process of bootstrap percolation, is attached to each graph to observe any spread in activity. An agent is assigned to each node, with all agents initially inactive and, from these, 20 nodes are randomly selected to form the initial active seed set, A_o. The activation mechanism involves an activation threshold r whereby an inactive node becomes active if it has at least r active neighbours. The recovery mechanism involves selecting a percentage of the active nodes for deactivation.

At each time step, the activation mechanism is applied synchronously to all nodes in the graph followed by the recovery mechanism applied to all active nodes. This cycle of activation and recovery at each time step is repeated until equilibrium; the number of active nodes is recorded at this point. Our experiments are then repeated at this Recovery Rate varying the activation threshold r from 2 to 10. We then repeat the experiments incrementing the recovery rate by 10% for each set, from 10 to 90%.

Our test simulations showed that equilibrium was reached rapidly for all graphs, and we therefore set an arbitrary cut-off point at 100 time steps, to ensure we captured equilibrium. In the case of graphs which had complete percolation in the standard bootstrap process, equilibrium cycled between full activation and deactivation of the recovery rate fraction.

Random Recovery

After each activation time step, the recovery rate percentage of active nodes is randomly selected to become inactive.

Targeted Recovery Based on Node Degree Ranking

After each activation time step, active nodes were ranked according to node degree, from highest to lowest, the recovery rate percentage of the top-ranked active nodes is then selected to become inactive.

5 Results

In this section we describe our results from simulating the processes of bootstrap percolation and bootstrap percolation with random recovery and then targeted recovery, all on the same set of hyperbolic random geometric graphs, using averaged data from the 20 graphs created at each parameter R.

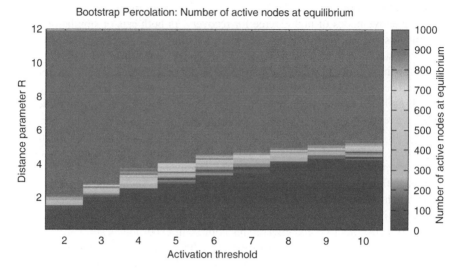

Fig. 4 Heat Map for $A_o = 20$, with AT from 2 to 10, and $R = 0.1$ to 12, showing the number of final active nodes at equilibrium

5.1 Bootstrap Percolation

The heatmap in Fig. 4 shows a distinct threshold above which the activity completely percolated and below which the activity failed to percolate. This percolation threshold lay between values of R from 1.5 to 5.8, representing edge densities of 0.005–0.1, respectively, demonstrating increasing edge density with each increase in activation value. Each increase in activation threshold has an inhibiting effect on the spread of activity, as it requires more of a node's contacts to be active before the activity will spread. Increasing the edge density allows an inactive node greater potential access to activated contacts, and this is manifested by the increase in the percolation threshold.

In all of our simulations, the heat maps displayed a clear percolation threshold. For comparison purposes, we developed a simple algorithm to find a representative point within the threshold, to allow for representation of the percolation threshold as a curve.

5.2 Bootstrap Percolation with Recovery

In the following charts, the standard Bootstrap Percolation process is depicted as a case of 0% recovery. Each fence plot has a differing recovery rate percentage ($RR\%$) and represents the percolation threshold for that particular recovery rate.

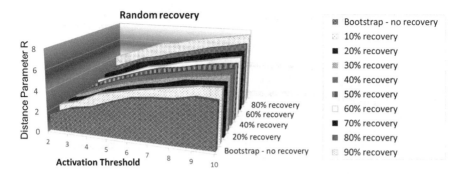

Fig. 5 Bootstrap percolation with RANDOM recovery

It can be clearly seen that either form of recovery significantly raises the threshold for complete percolation of the activity and that, in general, each increase in $RR\%$ increases this stepwise.

Bootstrap Percolation with Random Recovery

Figure 5 shows bootstrap percolation with recovery randomly targeting a percentage of active nodes following activation in each time step. It can be seen that, overall, the edge density of the threshold increases as the recovery rate increases from 10 to 90%, with the form of each curve similar to its neighbour. However, after the initial increase at 10%, increasing the recovery rate had little further effect before 50% recovery.

We can also gain information by looking at the edges of each curve. Percolation threshold values arising from an activation threshold of 2 are depicted on the left wall of the fence plot, whereas percolation threshold values arising from increasing the activation threshold to 10 are depicted on the right edge; the increase in activation threshold means that more active contacts are required for an inactive node to become activated. For an activation threshold of 2, these values for R at the percolation threshold increased from 2 to 4.7, representing edge densities ranging from 0.01 to 0.055. For an activation threshold of 10, the values for R at the percolation threshold increased from 5.2 to 8, representing edge densities from 0.075 to 0.3.

Bootstrap Percolation with Targeted Recovery

Figure 6 shows bootstrap percolation with recovery selectively targeting the top-ranked active nodes of the highest degree after each activation period. This clearly shows that the corresponding edge density of thresholds has increased significantly for all parameters, when compared with randomly selected recovery. Unlike random recovery, where the increase in recovery percentage had no effect between 10

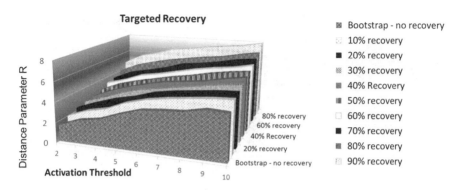

Fig. 6 Bootstrap percolation with TARGETED recovery

and 50%, the percolation threshold curves for targeted recovery showed a regular stepwise increase as the recovery percentage increased, for all percentages.

The most notable observation, with each percentage increase in recovery rate, was that for an activation threshold of 2, the percolation threshold occurred at values of R between 2 and 4, representing edge densities in the range 0.01–0.037, and that for an activation threshold of 10, the values for R at the percolation threshold increased from 5.2 to 7.6, representing edge densities of 0.075–0.25, with the form of each curve again similar to its neighbour. This demonstrates that selectively targeting the nodes of the highest degree for recovery has had a significantly greater impact on the spread of the activity compared with random selection, and particularly when compared with the standard bootstrap process.

6 Conclusion

We have studied bootstrap percolation on hyperbolic geometric graphs and noticed that there was a clear threshold above which the activity percolated to all nodes and below which the activity failed to percolate. We developed a modified form of bootstrap percolation which allowed for recovery of active nodes to inactive state. In our experiments with recovery on the same set of graphs, we noticed that with random recovery, after the initial increase at 10%, there was little further impact until we increased the recovery rate to 50%, at which the percolation threshold was significantly increased to graphs with higher edge density, with all curves of similar form to their neighbours.

Our experiments with targeted recovery, based on the top-ranked nodes of the highest degree, showed that this delayed the threshold even further, and had a stepwise increase in the threshold for all recovery rate increases. This suggests that nodes of high degree are influential in the bootstrap percolation process. Following these results, we surmise that it may be possible to fine-tune our targeting, by choosing differing node properties, or by targeting nodes highly ranked for a combination of properties.

Our next intention is to target active nodes for recovery based on properties which are significantly skewed in the hyperbolic graphs, such as clustering coefficient, betweenness centralisation and closeness centralisation. Our intuition is that finding measures which are highly skewed in the hyperbolic graphs will allow us to significantly delay percolation.

It would also be interesting to specifically target the most influential nodes in the network, with a priori rankings, so that whenever such a node became active it would always revert to inactivity. This would effectively immunise that node from activity and would represent an influential player constantly resisting an onslaught of activity.

Another potential area of study would concentrate on graphs within the percolation threshold zone of transition, investigating graph or individual node properties that facilitate the process above a certain edge density, or those which impede percolation below a certain edge density. It might be useful to observe the impact of targeted recovery within this percolation threshold in order to determine the local structural properties of individual nodes, or neighbourhoods, which might have the greatest impact on global outcomes in the graph.

References

1. Albert, R., Jeong, H., Barabási, A.L.: Error and attack tolerance of complex networks. Nature **406**(6794), 378–382 (2000)
2. Amini, H., Fountoulakis, N.: Bootstrap percolation in power-law random graphs. J. Stat. Phys. **155**(1), 72–92 (2014)
3. Balister, P., Bollobàs, B., Johnson, J.R., Walters, M.: Random majority percolation. Random Struct. Algoritm. **36**(3), 315–340 (2010)
4. Balogh, J., Pittel, B.G.: Bootstrap percolation on the random regular graph. Random Struct. Algoritm. **30**(12), 257–286 (2007)
5. Barabási, A.L.: Network Science. Cambridge University Press, Cambridge (2016)
6. Baxter, G.J., Dorogovtsev, S.N., Goltsev, A.V., Mendes, J.F.: Bootstrap percolation on complex networks. Phys. Rev. E **82**(1), 011103 (2010)
7. Bénézit, F., Dimakis, A.G., Thiran, P., Vetterli, M.: Order-optimal consensus through randomized path averaging. IEEE Trans. Inf. Theory **56**(10), 5150–5167 (2010)
8. Bringmann, K., Keusch, R., Lengler, J.: Geometric inhomogeneous random graphs. Preprint (2015). arXiv:1511.00576
9. Bullmore, E., Bassett, D.: Brain graphs: graphical models of the human brain connectome. Annu. Rev. Clin. Psychol. **7**, 113–140 (2011)
10. Candellero, E., Fountoulakis, N.: Clustering and the hyperbolic geometry of complex networks. In: Bonato, A., Graham, F., Pralat, P. (eds.) Algorithms and Models for the Web Graph. WAW 2014. Lecture Notes in Computer Science (including subseries Lecture Notes in Artificial Intelligence and Lecture Notes in Bioinformatics), vol. 8882, pp. 1–12. Springer, Cham (2014)
11. Candellero, E., Fountoulakis, N.: Bootstrap percolation and the geometry of complex networks. Stoch. Process. Appl. **126**, 234–264 (2015)
12. Centola, D.: The spread of behavior in an online social network experiment. Science **329**(5996), 1194–1197 (2010)
13. Chalupa, J., Leath, P.L., Reich, G.R.: Bootstrap percolation on a bethe lattice. J. Phys. C Solid State Phys. **12**(1), L31 (1979)

14. Coker, T., Gunderson, K.: A sharp threshold for a modified bootstrap percolation with recovery. J. Stat. Phys. **157**(3), 531–570 (2014)

15. Domingos, P., Richardson, M.: Mining the network value of customers. In: Proceedings of the Seventh ACM SIGKDD International Conference on Knowledge Discovery and Data Mining, KDD '01, pp. 57–66. ACM, New York (2001)

16. Gleeson, J.P.: Cascades on correlated and modular random networks. Phys. Rev. E **77**(4), 046117 (2008)

17. Gomez Rodriguez, M., Leskovec, J., Krause, A.: Inferring networks of diffusion and influence. In: Proceedings of the 16th ACM SIGKDD International Conference on Knowledge Discovery and Data Mining, pp. 1019–1028. ACM, New York (2010)

18. Jackson, M.O., López-Pintado, D.: Diffusion and contagion in networks with heterogeneous agents and homophily. Netw. Sci. **1**(01), 49–67 (2013)

19. Janson, S., Łuczak, T., Turova, T., Vallier, T.: Bootstrap percolation on the random graph $g_{n,p}$. Ann. Appl. Probab. **22**(5), 1989–2047 (2012)

20. Kempe, D., Kleinberg, J.M., Tardos, É.: Influential nodes in a diffusion model for social networks. In: ICALP, vol. 5, pp. 1127–1138. Springer, Berlin (2005)

21. Kempe, D., Kleinberg, J.M., Tardos, É.: Maximizing the spread of influence through a social network. Theory Comput. **11**(4), 105–147 (2015)

22. Krioukov, D., Papadopoulos, F., Kitsak, M., Vahdat, A., Boguñá, M.: Hyperbolic geometry of complex networks. Phys. Rev. E **82**, 036106 (2010)

23. Leskovec, J., Backstrom, L., Kleinberg, J.: Meme-tracking and the dynamics of the news cycle. In: Proceedings of the 15th ACM SIGKDD International Conference on Knowledge Discovery and Data Mining, pp. 497–506. ACM, New York (2009)

24. Liben-Nowell, D., Kleinberg, J.: Tracing information flow on a global scale using internet chain-letter data. Proc. Natl. Acad. Sci. **105**(12), 4633–4638 (2008)

25. Myers, S.A., Zhu, C., Leskovec, J.: Information diffusion and external influence in networks. In: Proceedings of the 18th ACM SIGKDD International Conference on Knowledge Discovery and Data Mining, pp. 33–41. ACM, New York (2012)

26. Papadopoulos, F., Psomas, C., Krioukov, D.: Network mapping by replaying hyperbolic growth. IEEE/ACM Trans. Networking **23**(1), 198–211 (2015)

27. Pastor-Satorras, R., Vespignani, A.: Epidemic spreading in scale-free networks. Phys. Rev. Lett. **86**(14), 3200 (2001)

28. Pržulj, N.: Biological network comparison using graphlet degree distribution. Bioinformatics **23**(2), e177–e183 (2007)

29. Rocchini, C.: Order-3 heptakis heptagonal tiling. https://commons.wikimedia.org/wiki/File: Order-3_heptakis_heptagonal_tiling.png (2007). Accessed 15 May 2017

30. Sahini, M., Sahimi, M.: Applications of Percolation Theory. CRC Press, Boca Raton (1994)

31. Shrestha, M., Moore, C.: Message-passing approach for threshold models of behavior in networks. Phys. Rev. E **89**(2), 022805 (2014)

32. Tassier, T.: Simple epidemics and SIS models. In: The Economics of Epidemiology, pp. 9–16. Springer, Berlin (2013)

33. von Looz, M., Staudt, C.L., Meyerhenke, H., Prutkin, R.: Fast generation of dynamic complex networks with underlying hyperbolic geometry. Preprint (2015). arXiv:1501.03545

34. Watts, D.J.: A simple model of global cascades on random networks. Proc. Natl. Acad. Sci. **99**(9), 5766–5771 (2002)

Process-Driven Betweenness Centrality Measures

Mareike Bockholt and Katharina A. Zweig

Abstract In network analysis, it is often desired to determine the most central node of a network, for example, for identifying the most influential individual in a social network. Borgatti states that almost all centrality measures assume that there exists a process moving through the network from node to node (Borgatti, Soc Netw 27(1):55–71, 2005). A node is then considered as central if it is important with respect to the underlying process. One often used measure is the betweenness centrality which is supposed to measure to which extent a node is "between" all other nodes by counting on how many shortest paths a node lies. However, most centrality indices make implicit assumptions about the underlying process. However, data containing a network and trajectories that a process takes on this network are available: this can be used for computing the centrality. Hence, in this work, we use existing data sets, human paths through the Wikipedia network, human solutions of a game in the game's state space, and passengers' travels between US American airports, in order to (1) test the assumptions of the betweenness centrality for these processes, and (2) derive several variants of a "process-driven betweenness centrality" using information about the network process. The comparison of the resulting node rankings yields that there are nodes which are stable with respect to their ranking while others increase or decrease in importance dramatically.

Keywords Network analysis · Centrality measures · Network processes · Path data analysis

M. Bockholt (✉) · K. A. Zweig
Department of Computer Science, University of Kaiserslautern, Kaiserslautern, Germany
e-mail: mareike.bockholt@cs.uni-kl.de; zweig@cs.uni-kl.de

© Springer International Publishing AG, part of Springer Nature 2018
R. Alhajj et al. (eds.), *Network Intelligence Meets User Centered Social Media Networks*, Lecture Notes in Social Networks,
https://doi.org/10.1007/978-3-319-90312-5_2

17

1 Introduction

An often performed task in network analysis is the identification of the most important nodes in a network. The goal might be to find the most influential individual in a social network, the most vulnerable location in a transportation network, or the leader in a terrorist network [6, 17]. The identification of such nodes is usually done with a centrality measure which computes a value for each node of the network based on the network structure [2, 10, 20]. The concept of centrality in networks was first introduced by Bavelas in the late 1940s who considered human communication networks [2]. Inspired by this idea, a large number of different methods for measuring the centrality of a node were proposed in the following decades, where the best known centrality measures are degree centrality [9], closeness centrality [9], betweenness centrality [1, 9], and Eigenvector centrality [3] (for an overview, cf. [5] or [14]).

An important contribution was made by Borgatti who states that almost all centrality indices are based on the assumption that there is some kind of traffic (or communication or process) flowing through the network by moving from node to node [4]. This might be the propagation of information in a social network, packages being routed through the WWW, or the spreading of a disease in human interaction networks. A node is then considered as central, that is it is assigned a high value of the measure, if it is somehow important with respect to this underlying process. However, different centrality measures make different assumptions about the process flowing through the network. Some measures are based on shortest paths in the network, assuming that the underlying process moves on shortest paths. Others assume that, if whatever flows through the network is at one node, it will spread simultaneously to all the node's neighbors, while others assume that it can only be at one node at a time.

Borgatti provides a typology of the most popular centrality measures by considering the network process [4] and argues that centrality measure values for a network can only be interpreted in a meaningful way, if the assumptions of the measure respect the properties of the process. He identifies two different dimensions by which the process can vary: First, on which type of trajectory does the process move through the network, and second, how does the process spread from node to node? For the first dimension, he differentiates between shortest paths, paths (not necessarily shortest, but nodes and edges can only occur at most once in it), trails (edges might occur several times in it, while nodes cannot be repeated), and walks (in which nodes and edges might occur several times). The second identified dimension is the "mechanism of node-to-node transmission": A process might *transfer* a good from node to node (e.g. a package), or it passes something to the next node by *duplicating* it: whatever flows through the network is passed to the next node and simultaneously stays at the current node (e.g. information flowing through a network or a viral infection: it will still be at the current node after it was passed to another node). Duplication can take place in a serial (one neighbour at once) or in a

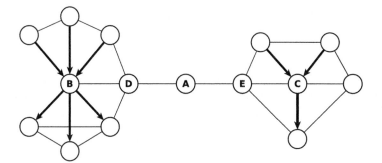

Fig. 1 Introductory example. Node A can be seen as a gatekeeper between the two node groups. Classic betweenness centrality hence assigns a high value to A. If, however, the network process only moves within the two groups (indicated by boldly drawn edges), should A still be considered as the most central node of the network?

parallel manner (simultaneously to all neighbours). This categorization is a helpful tool for choosing an appropriate centrality measure, given a network and a process on the network.

The betweenness centrality is an often used centrality measure in social network analysis [1, 9]. The idea is to measure to which extent a node v is positioned "between" all other nodes. A node with a high betweenness value is assumed to have control over other nodes: by passing information or withholding it, it can control the information flow. Consider for example the graph in Fig. 1 in which node A can be seen as the gatekeeper between the left subgraph and the right subgraph: A can prevent information being passed to the other subgraph because all paths between the two subgraphs pass through A. Betweenness centrality counts how many shortest paths from any node s to any other node t pass through the given node v and averages over all node pairs (s, t). In order to account for cases in which there are many shortest paths between s and t, but only a few of them pass through v, the measure normalizes by the total number of shortest paths between s and t. Formally, with σ_{st} denoting the number of shortest paths from a node s to a node t (with $\sigma_{st} = 1$ if $s = t$) and with $\sigma_{st}(v)$ denoting the number of shortest paths from s to t that pass through v, we can define the betweenness centrality for a node v as[1]

$$B_c(v) = \sum_{\substack{s \in V, \\ s \neq v}} \sum_{\substack{t \in V, \\ s \neq t \neq v}} \frac{\sigma_{st}(v)}{\sigma_{st}} \qquad (1)$$

According to Borgatti [4], this measure is appropriate for networks on which the process of interest literally *moves* from node to node by a transfer mechanism and travels on shortest paths. This becomes obvious in Fig. 1: if there were added two

---

[1]In literature, it is often noted as C^B. Since we are only considering this centrality measure and for a better readability in Sect. 6, we use this notation.

additional nodes D' and E', and edges linking D to D', D' to E', and E' to E, the high importance of A as a gatekeeper can only be still justified if the process exclusively takes shortest paths and can only take *one* path at once.

When taking a closer look at the formula, it is easy to see that there are even more simplifying assumptions in this measure: it assumed that the process flows between any node pair (s, t) and the amount of traffic flowing from s to t is equal (and equally important) for all node pairs. For a given network with a process fulfilling the two main assumptions, these further assumptions are justifiable: If there is no other information about the process available, these assumptions are the best possible. However, if the information of how the process moves through the network is available, the importance of the nodes can be measured differently. Consider again Fig. 1 in which the bold directed edges indicate that there is communication between those nodes and no communication between the other nodes. Hence, in both groups, there are many entities moving from one of the upper nodes to (the) one node below, using the nodes B and C, respectively, but no entity moving from one group to the other. Although the assumptions of moving on shortest paths and by a transfer mechanism are met, should the node A still be considered as the most central node if there is actually *no* communication at all between those two groups?

However, there are data sets available containing a network structure and trajectories that the network's process has taken. It is therefore possible to actually use the information of how the process moves through the network in order to assign a centrality value to the nodes. It is clear that this is a different approach than the classic betweenness centrality or other existing centrality measures: classic centrality indices solely use the *structure* of the given graph in order to compute centrality values for the nodes—the choice which centrality index is appropriate for this network can then be taken based on the knowledge about the properties of the process. Here, the information about the process is already used for *computing* the centrality value.

Hence, the contribution of this work is the following: we use three data sets in order to (1) investigate whether the assumptions of the betweenness centrality are met by those processes, and (2) incorporate the information about the process contained in the data sets into a *process-driven betweenness centrality* (PDBC). Several variants of PDBC measures are introduced in order to analyse which piece of information affects the node ranking in which manner. For example, one variant only counts the shortest paths between those node pairs which are source and destination of the process at least once. Another variant does not count the number of shortest paths but counts the number of actually used trajectories in which a node is contained. Those variants are then applied on the available data sets and the resulting node rankings of each variant compared to each other. It can be observed that the resulting rankings show a high correlation to each other, but there are nodes whose importance increases or decreases considerably.

This article is therefore structured as follows: Sect. 2 introduces the necessary definitions and notations before Sect. 3 presents related work in this area. Section 4 gives a description of the available data sets. Section 5 discusses the assumptions of the betweenness centrality and tests whether those assumptions are met in the data sets. Section 6 will introduce four variants of a process-driven betweenness centrality and Sect. 7 discusses the results of the PDBC measures on the given data sets, before Sect. 8 summarises the articles and gives an outlook for future work.

2 Definitions

$G = (V, E)$ denotes a directed simple unweighted graph with vertex set V, and edge set $E \subseteq V \times V$. A *path* is an alternating (finite) sequence of nodes and edges, $P = (v_1, e_1, v_2, \ldots, e_{k-1}, v_k)$ with $v_i \in V$ and $e_j = (v_j, v_{j+1}) \in E$ for all $i \in \{1, \ldots, k\}$ and $j \in \{1, \ldots, k - 1\}$, respectively.[2] Since G is simple, P is uniquely determined by its node sequence and the notation can be simplified to $P = (v_1, v_2, \ldots, v_k)$. The length of a path P ($|P|$) is defined as its number of (not necessarily distinct) edges. The start node of the path P is denoted by $s(P) = v_0$, the end node by $t(P) = v_k$ (occasionally referred to as source/start and destination/target). Let $d(v, w)$ denote the length of the shortest path from node v to node w. If w cannot be reached from v, we set $d(v, w) := \infty$.

3 Related Work

As illustrated in Sect. 1, the betweennness centrality contains several assumptions. Since these assumptions are not met in all networks and all processes, there have been proposed many variants of the betweenness centrality. The most prominent variants questioning the assumption of shortest paths are the flow betweenness centrality of Freeman [11], and the random walk betweenness centrality by Newman [16], but there are also variants as routing betweenness [7] or variants for dynamic networks where paths up to a certain factor longer than the shortest one contribute to the centrality [6]. A variant questioning the assumption that all paths contribute equally to centrality value is the length-scaled betweenness [5].

Path Analysis The idea of analysing the process moving through a network is not new. In many real-world networks, entities use the network to navigate through it by moving from one node to the other. This includes Internet users surfing the Web yielding clickstream data, or passengers travelling through a transportation network. The navigation of an entity in a network to a target node, but with only local information about the network structure is called *decentralized search*. The fact that humans are often able to find surprisingly short paths through a network was already illustrated by Milgram in his famous small-world experiment in 1967 [15]. An answer of *how* humans actually find these short paths was not known before

---

[2]Note that we do not require the nodes and edges to be pairwise distinct. In some literature, P would be referred to as a walk.

Kleinberg investigated which effect the network structure has on the performance of any decentralized search algorithm [13]. West and Leskovec analysed the navigation of information seekers in the Wikipedia networks [23] and also find that human paths are surprisingly short, and show similar characteristics in their structure. This is also found by Iyengar et al. [21] who considered human navigation in a word game.

Combination of Centrality and Path Analysis There are approaches of using the information about the network process in order to infer knowledge about the network itself: West et al. [24] use the paths of information seekers in the Wikipedia network to compute a semantic similarity of the contained articles. Rosvall et al. [19] show how real-word pathway data incorporated into a second-order Markov model has an effect on the node rankings in an approach generalising PageRank, and can derive communities of the network by the network's usage patterns. Zheng [26] identify popular places based on GPS sequences of travellers. Also based on GPS trajectories, the approach of Yuan et al. [25] makes use of Taxi drivers' trajectories in order to compute the effectively quickest path between two places in a city. Dorn et al. [8] can show that the results of betweenness centrality in the air transportation network significantly change if the centrality considers the number of actually taken paths traversing through an airport instead of all possible shortest paths.

4 Data Sets

The goal is to investigate whether the assumptions built into the betweenness centrality are actually met by a data set containing the information how a process moves in a network. Since we consider the betweenness centrality, it is, according to Borgatti [4], only meaningful to consider processes which (1) move by a transfer mechanism, and (2) move through the network with a target, that is a predetermined node to reach. The first condition implies that the movement of the entity can be modelled as a path. Data sets appropriate to use in this work, hence, need to fulfil the following requirements:

1. It contains a network structure, given as graph $G = (V, E)$
2. It contains a set of paths of one or several entities moving through the network, given as $\mathcal{P} = \{P_1, \ldots, P_\ell\}$
3. The entities move through the network with a predetermined goal to reach
4. That they aim to reach as soon as possible

We use the following data sets (see also Table 1).

Wikispeedia This data set provided by West et al. [23, 24] contains (a subset of) the network of Wikipedia articles where a node represents an article and there exists an edge from one node to another if there exists a link in the one article leading to the other article. The paths represent human navigation paths, collected by the game *Wikispeedia* in which a player navigates from one (given) article to another (given)

Table 1 Overview of the used data sets

Data set	Source	Graph type	Nodes	Edges	
Wikispeedia	[23, 24]	Directed	Articles	Hyperlinks	
Rush Hour	[12]	Undirected	Configurations	Valid game moves	
DB1B	[18]	Directed	Cities	Non-stop airline connections	
				Path length	
Data set	$|V|$	$|E|$	$|\mathcal{P}|$	Range	Average
Wikispeedia	4592	119,804	51,306	[1, 404]	5
Rush Hour	364	1524	3044	[3, 33]	5
DB1B	419	12,015	63,681,979	[1, 14]	1.3

article by following the links in the articles. Only paths reaching their determined target are considered. We model the Wikispeedia network as a directed, simple unweighted graph (G_W) and the set of paths as $\mathcal{P}_W$.

Rush Hour This data set contains a state space of a single-player board game called *Rush Hour* where each node represents a possible game configuration and there is an edge from node v to node w if configuration w can be reached from v by a valid move in the game. We only include those nodes which are reachable from the node representing the start configuration of the game. Since all moves are reversible, the network is modelled as undirected unweighted graph (denoted by G_R) where one node represents the start configuration and one or more nodes represent configurations in which the game is solved (final nodes). A path in this data set is then the solution of a player trying to reach the final node from the start node by a sequence of valid moves. Only paths ending in a final node are considered. The set of paths will be denoted by $\mathcal{P}_R$. The data set was collected by Pelánek and Jarušek [12] by their web-based tool for education (available under tutor.fi.muni.cz).

DB1B This data set contains a sample of 10% of all airline tickets of all reporting airlines, including all intermediate stops of a passenger's travel, provided by the US Bureau of Transportation Statistics which publishes for each quarter of a year the Airline Origin and Destination Survey (DB1B) [18]. We consider the passengers' travels for the years 2010 and 2011. A path is a journey of a passenger travelling from one airport to another, possibly with one or more intermediate stops. The network (G_D) is modelled as a simple directed and unweighted graph and is extracted from the ticket data by identifying city areas (possibly including more than one airport) as nodes, and adding a directed edge from a node v to a node w if at least one passenger's journey contains a flight from one airport in the area of node v to another airport in the area of node w. Passengers' journeys which are symmetric in the sense that the passenger travels from airport A to airport B over i intermediate airports and via the same intermediate airports back to A will be considered as two paths: one from A to B and one from B to A.

5   Assumptions of Betweenness Centrality

We chose our data sets such that the main assumptions of the betweenness centrality are met: A process is flowing through the network by transfer mechanism, and whatever flows through it has a target to reach. Section 1 already pointed out that there are more assumptions. The next section will investigate whether those assumptions are satisfied in the selected data sets.

Process Moves on Shortest Paths For each G_X and corresponding $\mathcal{P}_X$, $X \in \{W, D, R\}$, and for each $P \in \mathcal{P}_X$, we compare $|P|$ with $d(s(P), t(P))$. The results are shown in Fig. 2a. For Rush Hour, $d(s(P), t(P)) = 3$ for each $P \in \mathcal{P}_R$ because we only consider paths reaching a final node from the start node—which is possible within three moves. For Wikispeedia, $1 \leq d(s(P), t(P)) \leq 6$ for all $P \in \mathcal{P}_W$. One outlier path of length 404 and two paths of length 101 and 104 have been taken out of the analysis. All node pairs in the DB1B network that were source and destination of any of the paths in $\mathcal{P}_D$ can be reached within four flights. We can observe that for the DB1B data set, the taken journey is almost always the shortest path between the airports. For Wikispeedia, the variation of the path lengths is much larger than for DB1B. Although the median length of the paths is greater than the length of the shortest path between its start and end nodes, the difference is only about 1. The same holds for Rush Hour. Hence, for all of the considered data sets, the assumption that the process is moving on shortest paths is approximately true.

Process Moves Between Every Node Pair Since the betweenness centrality counts the (fraction of) shortest paths including a given node v and those fractions are summed up over all possible node pairs (s, t) (with $s \neq v$ and $s \neq t \neq v$), it is assumed that there is traffic between every pair of nodes. Table 2 shows whether this is a valid assumption for our data sets: In all data sets, almost all nodes are visited by at least one path. Since the network for DB1B was constructed from the paths, the number of used nodes is equal to the total number of nodes. It is not surprising that the fraction of used nodes in the Rush Hour data set is much smaller than in the other data sets, since all paths start at the network's start node and end in one of the final nodes. In the Wikispeedia data set, 90% of all nodes were used by at least one path. If, however, a node is only counted if at least ten paths include this node, the percentage drops to 74%. Figure 2b shows this fact: It shows which portion of the nodes (contained in at least one path) is used by at least which portion of paths.

Although the majority of the nodes are included in at least one path of the data set, only a small fraction of the node *pairs* is actually used as source and destination of any path (cf. Table 2 and Fig. 2c). For Rush Hour and Wikispeedia, it rather seems to be an artifact: out of 132,132 node pairs in G_R, only 20 pairs can actually be used by a valid solving path, since it must start in the start configuration and end in one of the 20 final configurations. In G_W, there are more than 21 million node pairs, but the data set contains "only" about 50*k* paths which explains the small value. For G_D on the other side, the fraction of used node pairs is, although the highest of

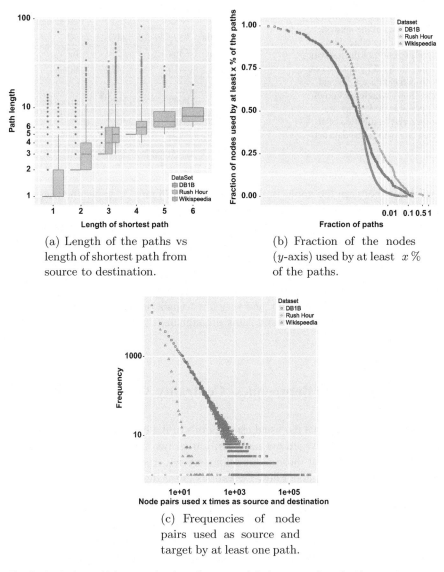

(a) Length of the paths vs length of shortest path from source to destination.

(b) Fraction of the nodes (y-axis) used by at least $x\,\%$ of the paths.

(c) Frequencies of node pairs used as source and target by at least one path.

Fig. 2 Analysis to which extent the given data sets satisfy the assumptions that the process moves on shortest paths (**a**), and that the process moves between each pair of nodes with the same intensity (**c**)

all data sets, surprisingly small. Although the data contain $63m$ passengers' travels over a time period of 2 years, almost half of all airport (city) pairs is never taken as a passenger's journey.

Equal Amount of Flow Between Every Pair of Nodes While the previous paragraph showed that the assumption there is flow between every pair of nodes in the network is not met in the data sets, this paragraph provides evidence that the assumption of

Table 2 Usage frequencies of the nodes of the networks: first part of the table shows which fraction of the network nodes are used by at least one path in the data set. Second part shows which fraction of all possible node pairs are source and destination of at least one path in the data set

Data set	Nodes	Used	Percentage (%)
Wikspeedia	4592	4166	90
Rush Hour	364	231	63
DB1B	419	419	100

Data set	Pairs	Used	Percentage (%)
Wikspeedia	21,081,872	28,706	0.14
Rush Hour	132,132	19	0.01
DB1B	175,142	92,242	52.7

equal flow between the nodes is also not true here. Figure 2c shows the frequency distribution of all node pairs which are the source and destination of at least one path in the data set, that is how many node pairs (y-axis) are used by exactly x paths in the data set as source and destination. Note the logarithmic scale on both axes. We see that although on different magnitudes, the behaviour is qualitatively the same for all data sets. When picking a node pair uniformly at random out of all node pairs used at least once, the probability is very high to pick a pair which is the source and destination of only one path, and very low to pick one which is the source and destination of more than a thousand paths although there are more than $50k$ and $63m$ paths, respectively.

6   Process-Driven Betweenness Centrality Measures

Since we have shown that the usual assumptions of the betweenness centrality and about its underlying flow are not necessarily true, two questions arise. First, how much does it matter that those assumptions are not true, and second, can we do better? The information about how the entities move through the network is given in the data sets and can therefore be incorporated into a *process-driven betweenness centrality measure* (PDBC) which does not rely on the assumptions of shortest paths, flow between every node pair with equal intensity. The following section introduces four PDBC variants using different pieces of information contained in the data sets. The first two keep the assumption of using shortest paths (indicated by subscripted S), the next two count *real* paths from $\mathcal{P}$ (indicated by a subscripted R). All four require a graph $G = (V, E)$ and a set of paths $\mathcal{P}$ in G to be given. As a general framework, we introduce a weighted betweenness centrality by

$$B_w(v) = \sum_{s \in V} \sum_{t \in V} w(s, t, v) \cdot \frac{\sigma_{st}(v)}{\sigma_{st}} \tag{2}$$

with a weight function $w : V \times V \times V \to \mathbb{R}$. The standard betweenness centrality is then B_c with the weight function $w_c(s, t, v) = 0$, if $s = t$ or $s = v$ or $v = t$, and $w_c(s, t, v) = 1$ otherwise, and the standard betweenness centrality including endpoints is B_E with the weight function $w_E(s, t, v) = 1$ for all $s, t, v \in V$.

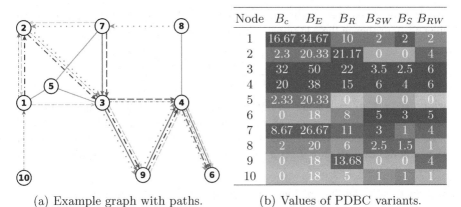

Node	B_c	B_E	B_R	B_{SW}	B_S	B_{RW}
1	16.67	34.67	10	2	2	2
2	2.3	20.33	21.17	0	0	4
3	32	50	22	3.5	2.5	6
4	20	38	15	6	4	6
5	2.33	20.33	0	0	0	0
6	0	18	8	5	3	5
7	8.67	26.67	11	3	1	4
8	2	20	6	2.5	1.5	1
9	0	18	13.68	0	0	4
10	0	18	5	1	1	1

(a) Example graph with paths.   (b) Values of PDBC variants.

Fig. 3 Example graph G and PDBC values for the shown graph

Figure 3a shows an example graph with four paths in it, indicated by the directed edges of different colours. Then, $\mathcal{P} = \{P_1, P_2, P_3, P_4, P_5, P_6\}$ with $P_1 = (10, 1, 2, 3, 4, 6)$, $P_2 = (7, 3, 4, 6)$, $P_3 = (8, 7, 2, 3, 9, 4, 6)$, $P_4 = (7, 3, 9, 4, 6)$, $P_5 = (1, 2, 3, 9, 4)$ and $P_6 = (7, 2, 3, 9, 4, 6)$. The table in Fig. 3b shows the centrality values of the nodes with respect to each measure.

Note that the evaluation of the results in Sect. 7 will focus on the resulting *rankings* of the nodes with respect to the different measure variants, and not the actual measure *values*; hence, no effort is made with any normalization of the measures.

Variant B_S Keeping the assumption of shortest paths, this variant only considers shortest paths between nodes being source and destination of at least one path in $\mathcal{P}$. We define

$$B_S(v) = \sum_{s \in V} \sum_{t \in V} w_S(s, t, v) \cdot \frac{\sigma_{st}(v)}{\sigma_{st}} \qquad (3)$$

with the weight function

$$w_S(s, t, v) = \begin{cases} 1 & \text{if} \exists P \in \mathcal{P} : s(P) = s \text{ and} t(P) = t \\ 0 & \text{else} \end{cases} \qquad (4)$$

For Fig. 3, the weights are $w_S(1, 4, \cdot) = w_S(7, 6, \cdot) = w_S(8, 6, \cdot) = w_S(10, 6, \cdot) = 1$ and $w_S(s, t, v) = 0$ for all other $s, t, v \in V$. For node 3, the centrality value is then $B_S(3) = \frac{\sigma_{1,4}(3)}{\sigma_{1,4}} + \frac{\sigma_{8,6}(3)}{\sigma_{8,6}} + \frac{\sigma_{10,6}(3)}{\sigma_{10,6}} + \frac{\sigma_{7,6}(3)}{\sigma_{7,6}} = \frac{1}{1} + \frac{0}{1} + \frac{1}{1} + \frac{1}{2} = 2.5$.

Variant B_{SW} Section 5 showed that in the considered data sets, there is much more communication between some node pairs than between others. If a node is contained in most of the paths between all highly demanded node pairs, the node should have a higher centrality than if it contained in the paths between less demanded node pairs.

We therefore make the weight function proportional to the amount of flow between the node pair (hence the additional subsripted W). Formally, we define

$$B_{SW}(v) = \sum_{s \in V} \sum_{t \in V} w_{SW}(s, t, v) \cdot \frac{\sigma_{st}(v)}{\sigma_{st}} \tag{5}$$

with $w_{SW}(s, t, v) = |\{P \in \mathcal{P} | s(P) = s, t(P) = t\}|$. For the example in Fig. 3a, this yields $w_{SW}(1, 4, \cdot) = w_{SW}(8, 6, \cdot) = w_{SW}(10, 6, \cdot) = 1, w_{SW}(7, 6, \cdot) = 3$ and $w_{SW} = 0$ in all other cases. Since shortest paths are counted, it is clear that often visited nodes as node 2 or 9 get a small value as they are not on any shortest path between the used nodes.

Variant B_R Unlike the previous two measures, this and the next variant count in how many *real* paths a node is contained in (therefore the subscript R). We define a *process-driven* version of σ_{st} and $\sigma_{st}(v)$: In order to keep the assumption that the process flows between any pair of nodes, we define $\sigma_{\cdot st \cdot}^{\mathcal{P}}$ as the number of paths in $\mathcal{P}$ *containing* s and t, and $\sigma_{\cdot st \cdot}^{\mathcal{P}}(v)$ as the number of paths in $\mathcal{P}$ that contain s and t, and v in between. (Otherwise, if the $\sigma_{\cdot st \cdot}^{\mathcal{P}}$ was defined as the number of paths in $\mathcal{P}$ with start node s and end node t, node pairs which are not start and end nodes of any path do not contribute to the centrality value.) We then define

$$B_R(v) = \sum_{s \in V} \sum_{t \in V} w_R(s, t, v) \cdot \frac{\sigma_{\cdot st \cdot}^{\mathcal{P}}(v)}{\sigma_{\cdot st \cdot}^{\mathcal{P}}} \tag{6}$$

with the convention $\frac{0}{0} = 0$ and $w_R(s, t, v) = 1$ for all $s, t, v \in V$ with $s \neq t$. Note that node pairs where at least one of the nodes is not contained in any paths in $\mathcal{P}$ do not contribute anything to the sum. In Fig. 3a, all nodes except of node 5 are contained in a path from $\mathcal{P}$, we therefore get for node 2,

$$B_R(2) = \frac{\sigma_{\cdot 1,2 \cdot}^{\mathcal{P}}(2)}{\sigma_{\cdot 1,2 \cdot}^{\mathcal{P}}} + \cdots + \frac{\sigma_{\cdot 7,9 \cdot}^{\mathcal{P}}(2)}{\sigma_{\cdot 7,9 \cdot}^{\mathcal{P}}} + \cdots + \frac{\sigma_{\cdot 10,6 \cdot}^{\mathcal{P}}(2)}{\sigma_{\cdot 10,6 \cdot}^{\mathcal{P}}} = \frac{2}{2} + \cdots + \frac{2}{3} + \cdots + \frac{1}{1} = 21.17 \tag{7}$$

We see that some nodes with a small centrality value with respect to B_c because they are not on (many) shortest paths have a larger centrality value in this variant. For example, node 9 or node 2 have a minor importance with respect to B_S but rise in importance wrt B_R because they are contained in a certain number of real paths.

Variant B_{RW} This variant combines all three kinds of information about the process in the network: counting real paths instead of shortest paths, only between those node pairs actually used as source and destination of the process instead of all, and weighting a node pair's contribution according to the amount of flow between the node pairs instead of assuming an equal amount. Formally, it is defined as

$$B_{RW}(v) = \sum_{s \in V} \sum_{t \in V} w_{RW}(s, t, v) \cdot \frac{\sigma_{st}^{\mathcal{P}}(v)}{\sigma_{st}^{\mathcal{P}}} = |\{P \in \mathcal{P} | v \in P\}| \tag{8}$$

Table 3 Categorization of the introduced process-driven betweenness centralities (PDBC)

	Count	How	Sum over s, t with	Weight
B_S	Shortest paths		$\exists P \in \mathcal{P} : s \to t$	1
B_{SW}	Shortest paths		$\exists P \in \mathcal{P} : s \to t$	# real paths $s \to t$
B_R	Real paths	$\to s \to v \to t \to$	$s, t \in V$	1
B_{RW}	Real paths	$s \to v \to t$	$\exists P \in \mathcal{P}: s \to t$	# real paths $s \to t$

with $w_{RW}(s, t, v) = w_{SW}(s, t, v) = |\{P \in \mathcal{P} | s(P) = s, t(P) = t\}|$. It is a kind of stress betweenness centrality. We see in Fig. 3 that node 3 which is central with respect to all previously considered centrality measures is also central with respect to B_{RW}, but nodes as 9 and 2 rise in importance because a considerable amount of paths passes through them.

Measures to Compare We have introduced four variants of PDBC measures (see also Table 3). Section 7 will describe how the centrality value of nodes (rather their position in the resulting ranking) will be affected when the PDBC measures are applied to the described data sets. For Wikispeedia and DB1B, it seems appropriate to compare the node rankings with those of B_E. However, for Rush Hour, this is different: Since G_R represents the state space of a game, a valid path in this graph needs to start in the start node of G_R, and, in order to be a solving path, needs to end in one of the final states. This is why we introduce the *game betweenness centrality* for this case by

$$B_G(v) = \sum_{s \in V} \sum_{t \in V} w_G(s, t, v) \cdot \frac{\sigma_{st}(v)}{\sigma_{st}} \tag{9}$$

with $w_G(s, t, \cdot) = 1$ if s is start node and t final node, and $w_G(s, t, \cdot) = 0$ otherwise.

7 Results

We computed the four PDBC variants and B_E for the networks described in Sect. 4. We are not interested in the exact values, but in the resulting order of importance of the nodes, only the resulting rankings are considered. We apply fractional ranking where nodes with the same measure value are assigned the average of ranks which they would have gotten without the ties.

Measure Correlations Figure 4 (left) shows the rankings of the nodes in the data sets wrt all PDBC measures as well as B_E and (for Rush Hour) B_G. We can observe that for Wikispeedia and DB1B, there is a strong correlation between the node rankings wrt all variants. However, the correlations are less strong for Wikispeedia than for DB1B. Furthermore, the correlation of the PDBC variants with B_E is very low for the Rush Hour data set (largest value is 0.46). However, the rankings wrt

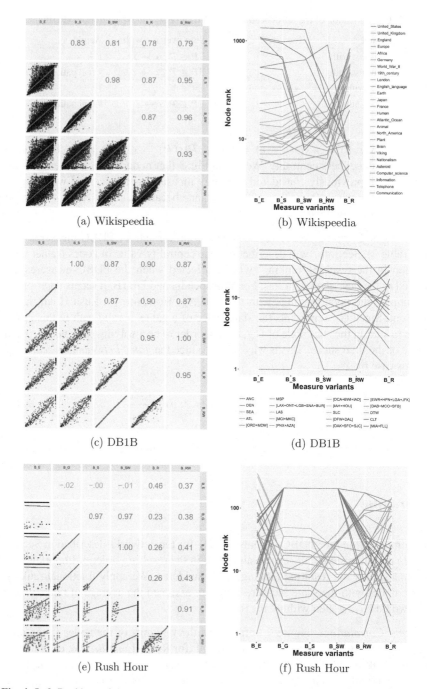

(a) Wikispeedia

(b) Wikispeedia

(c) DB1B

(d) DB1B

(e) Rush Hour

(f) Rush Hour

Fig. 4 Left Rankings of the network nodes in all PDBC variants and the standard betweenness centrality including endpoints (*BE*). Low value on the axes indicates high centrality value.

the PDBC variants counting shortest paths (B_{SW}, B_S) correlate with the game betweenness centrality (correlation of 0.97), but not those variants counting real paths (B_R, B_{RW}). In all three data sets, although most nodes have a similar ranking position in all variants, there are nodes which are rated as important by one measure and very unimportant by another measure.

Most Central Nodes We are interested in those nodes which are central with respect to at least one centrality measure. Figure 4 (right) shows the behaviour of the ranking of nodes which are among the ten nodes with highest centrality values for at least one of the measure variants. We can observe that in none of the data sets, it is the same set of ten nodes most central with respect to all measure variants. In all three data sets, the ranking wrt B_S is very similar to the "baseline" betweenness centrality, that is B_E for Wikispeedia and DB1B, and game betweenness centrality for Rush Hour. This implies that the PDBC variant with shortest paths and weights $\in \{0, 1\}$ does not affect the ranking of the most central nodes with respect to B_E. This is surprising because Sect. 5 showed that in those data sets, only a small part of all node pairs are actually used as source and target. Furthermore, for Rush Hour and DB1B, the ranking wrt B_{RW} is very different from the others, and for Wikispeedia, however, the ranking with respect to B_{RW} is not too different from the ones wrt B_{SW}, and B_S. It is remarkable that for Wikispeedia, there is one node (the article *United States*) which is the most central with respect to B_E, but also the most (or at most third) central with respect to all PDBC variants. This cannot be observed in any of the other two data sets. Also the nodes of the Wikispeedia network on ranking position 2 to 5 wrt B_E are among the most central ones wrt all PDBC variants (except of B_R). The scale by which the considered nodes increase or decrease in ranking positions is different for the data sets: for DB1B, all nodes among the ten most central nodes wrt at least one of the variants are among the 52 most central nodes wrt all variants (out of 419 possible ranking positions). This is different for the other two data sets: For Rush Hour, there are nodes which are among the ten most central nodes with respect to B_{SW} and B_{RW} but have a ranking of 357 (out of 364) with respect to B_E. However, when examining those nodes, it turns out that those are final nodes and nodes adjacent to final states. More interesting are those nodes which are among the least central nodes with respect to the game betweenness centrality, but among the ten most central nodes wrt B_R and B_{RW}—those measures which count actually used paths instead of shortest paths. Those nodes are neither start nor solution states, and although they are not on any shortest path from the start to any solution state, those nodes seem to be preferred by human players when solving

Fig. 4 (continued) The red lines are linear regressions of the nodes' rankings. The value of the (rounded) Pearson correlation coefficient between the corresponding rankings can be found in the corresponding box, it is red if $p < 0.05$, blue otherwise. **Right** Ranking behaviour of those nodes which are among the ten most central nodes with respect to at least one of the measures. Note the logarithmic scale on the y-axis. Colours are given according to the node's ranking wrt B_E (for Wikispeedia and DB1B) or wrt the game betweenness centrality (for Rush Hour)

this game. For Wikispeedia, there are nodes whose ranking increases considerably: from ranking position 1843 (out of 4592) wrt B_E to position 5 wrt B_R (node *Communication*) or from position 1206 to 6 wrt B_{SW} (*Telephone*). Those big jumps in rank position might, however, partly be an effect of the data collection: There are four source-target pairs which are suggested to the players with increased frequency as start and target nodes for the game for a certain period of time (*Pyramid*→*Bean*, *Brain*→*Telephone*, *Asteroid*→*Viking*, *Theatre*→*Zebra*) [22]. This means that paths with these sources/targets are contained more often than others. This explains why these nodes increase in importance when considering B_{SW} but does not explain why they increase in importance in the same magnitude with respect to B_R.

8   Summary and Future Work

This work used available data sets containing the trajectories of entities in a network structure in order to compute a process-driven centrality measure for the nodes. The idea is to use the available information about how a process moves through the network in order to rank the nodes according to their importance with respect to this process. We chose data sets such that the two main assumptions of the betweenness centrality are met and it is therefore appropriate to actually use this centrality measure. We could show that while assuming shortest paths in those data sets can be justified, other assumptions of the betweenness centrality are not satisfied. We therefore introduced four variants of process-driven betweenness centrality measures (PDBC) which incorporate the information contained in the data sets. The resulting node rankings of most variants show high correlations to each other as well as to the standard betweenness centrality. Nevertheless, we can observe that incorporating the information about the process has an effect on the most central nodes with respect to the standard betweenness centrality and can increase or decrease a node's ranking by several thousands of ranks. We furthermore observe that the different process-driven betweenness variants affect the node rankings in the three data sets in different ways.

Open questions left for future work are for example: it is obvious that the quality and validity of the results of PDBC is highly dependent on the quantity and quality of the available data. This yields two future directions: collect and compile more path data sets of high quality, and investigate which properties the set $\mathcal{P}$ must satisfy in order to make a reasonable statement. Additionally, networks develop dynamically over time: If an edge is not used by the process, it might disappear at some point, and if there is a large amount of traffic between two nodes with a larger distance, there might appear a shortcut between them. Considering the usage of the network by available path data might therefore be a tool for predicting a network's change of structure.

References

1. Anthonisse, J.M.: The rush in a directed graph. Technical Report BN 9/71, Stichting Mathematisch Centrum, Amsterdam (1971)
2. Bavelas, A.: A mathematical model for group structures. Hum. Organ. **7**(3), 16–30 (1948)
3. Bonacich, P.: Factoring and weighting approaches to status scores and clique identification. J. Math. Sociol. **2**(1), 113–120 (1972)
4. Borgatti, S.P.: Centrality and network flow. Soc. Netw. **27**(1), 55–71 (2005)
5. Borgatti, S.P., Everett, M.G.: A graph-theoretic perspective on centrality. Soc. Netw. **28**(4), 466–484 (2006)
6. Carpenter, T., Karakostas, G., Shallcross, D.: Practical issues and algorithms for analyzing terrorist networks. In: Proceedings of the Western Simulation MultiConference (2002)
7. Dolev, S., Elovici, Y., Puzis, R.: Routing betweenness centrality. J. ACM **57**(4), 25:1–25:27 (2010)
8. Dorn, I., Lindenblatt, A., Zweig, K.A.: The trilemma of network analysis. In: Proceedings of the International Conference on Advances in Social Networks Analysis and Mining (ASONAM), Washington, DC, pp. 9–14 (2012)
9. Freeman, L.C.: A set of measures of centrality based on betweenness. Sociometry **40**, 35–41 (1977)
10. Freeman, L.C.: Centrality in social networks conceptual clarification. Soc. Netw. **1**(3), 215–239 (1978)
11. Freeman, L.C., Borgatti, S.P., White, D.R.: Centrality in valued graphs: a measure of betweenness based on network flow. Soc. Netw. **13**(2), 141–154 (1991)
12. Jarušek, P., Pelánek, R.: Analysis of a simple model of problem solving times. In: Cerri, S., Clancey, W., Papadourakis, G., Panourgia, K. (eds.) Intelligent Tutoring Systems. Lecture Notes in Computer Science, vol. 7315, pp. 379–388. Springer, Berlin (2012)
13. Kleinberg, J.M.: Navigation in a small world. Nature **406**, 845–845 (2000)
14. Koschützki, D., Lehmann, K.A., Peeters, L., Richter, S., Tenfelde-Podehl, D., Zlotowski, O.: Centrality indices. In: Brandes, U., Erlebach, T. (eds.) Network Analysis. Lecture Notes in Computer Science, vol. 3418, pp. 16–61. Springer, Berlin (2005)
15. Milgram, S.: The small world problem. Psychol. Today **2**(1), 60–67 (1967)
16. Newman, M.E.: A measure of betweenness centrality based on random walks. Soc. Netw. **27**(1), 39–54 (2005)
17. Qin, J., Xu, J.J., Hu, D., Sageman, M., Chen, H.: Analyzing terrorist networks: a case study of the global salafi jihad network. Intell. Secur. Inform. **3495**, 287–304 (2005)
18. RITA TransStat: Origin and Destination Survey database (DB1B) (2016)
19. Rosvall, M., Esquivel, A.V., Lancichinetti, A., West, J.D., Lambiotte, R.: Memory in network flows and its effects on spreading dynamics and community detection. Nat. Commun. **5**, 4630 (2014)
20. Sabidussi, G.: The centrality index of a graph. Psychometrika **31**(4), 581–603 (1966)
21. Sudarshan Iyengar, S., Veni Madhavan, C., Zweig, K.A., Natarajan, A.: Understanding human navigation using network analysis. Top. Cogn. Sci. **4**(1), 121–134 (2012)
22. West, R.: Human navigation of information networks. Ph.D. thesis, Stanford University (2016)
23. West, R., Leskovec, J.: Human wayfinding in information networks. In: Proceedings of the 21st International Conference on World Wide Web, pp. 619–628. ACM, New York (2012)
24. West, R., Pineau, J., Precup, D.: Wikispeedia: an online game for inferring semantic distances between concepts. In: IJCAI International Joint Conference on Artificial Intelligence, pp. 1598–1603 (2009)
25. Yuan, J., Zheng, Y., Zhang, C., Xie, W., Xie, X., Sun, G., Huang, Y.: T-drive: driving directions based on taxi trajectories. In: Proceedings of the 18th SIGSPATIAL International Conference on Advances in Geographic Information Systems, pp. 99–108. ACM, New York (2010)
26. Zheng, Y., Zhang, L., Xie, X., Ma, W.Y.: Mining interesting locations and travel sequences from GPS trajectories. In: Proceedings of the 18th International Conference on World Wide Web, pp. 791–800. ACM, New York (2009)

Behavior-Based Relevance Estimation for Social Networks Interaction Relations

Pascal Held and Mitch Köhler

Abstract The problem of estimating the relevancy of an edge over the time a graph evolves is well known in the domain of social network analysis and can be used to make predictions about who will be friends as well as who is going to interact with each other in the future. Approaches incorporated in this prediction problem are mainly focusing on the amount or probability of interaction to compute an answer. Rather than characterizing an edge by the amount of interaction, we'd like to propose a method that introduces a behavioral component together with social decay to categorize whether a relationship's activity is rising, stable, or falling. Utilizing an individualized set of rules, one can focus on behavioral changes that are interesting to the object of investigation.

Keywords Evolving social graphs · Relevance Estimation · Interaction

1 Introduction

Estimating the relevancy of an edge is an important aspect in the research domain of *Link Prediction* which tries to predict who is going to interact with whom (again) in the future—based on past interaction behavior that can be modeled by a series of timed events where each event represents a recorded interaction between two nodes in a (perhaps social) graph.

Recent work in this field has shown that incorporating *recency of interactions* between nodes into prediction models yields accuracy improvements [4–6] and can be used to find cluster and communities in social networks [2].

P. Held (✉) · M. Köhler
Otto von Guericker University of Magdeburg, Institute for Intelligent Cooperating Systems, Magdeburg, Germany
e-mail: pascal.held@ovgu.de; mitch.koehler@st.ovgu.de
http://fuzzy.cs.ovgu.de

© Springer International Publishing AG, part of Springer Nature 2018   35
R. Alhajj et al. (eds.), *Network Intelligence Meets User Centered Social Media Networks*, Lecture Notes in Social Networks,
https://doi.org/10.1007/978-3-319-90312-5_3

In our work, we'd like to contribute an alternative approach to model the relevancy of *recent interaction history* based on *behavioral change* and *trend in interaction intensity*. Doing so, we see an edge between nodes as a boolean link (present or absent). Edges that are present carry an interaction history as well as a boolean flag for being currently considered as active (or relevant) with respect to the modeled problem domain. This differs from other authors' terminology where each edge represents exactly one interaction or a boolean link (i.e., a friendship) between nodes. Our approach only needs information inferred from the history of interaction—no neighborhood data will be incorporated.

One of our main motivations was to develop an easily parallelizable as well as memory and computational efficient relevance estimation of an edge between two nodes over time. Modeling the relevance of an *existing* interaction between nodes, we have to account for scenarios of *active* communication. However, what happens if a thriving communication *suddenly* stops and *no* more interaction events are exchanged between both nodes? We believe that the psychological impression of a sudden stop of a thriving communication can be different to a steadily declining communication intensity (and must be differently treated during relevancy estimation). Though we do not evaluate this belief, we introduced a component for *social decay* in our model that reflects this thought and its implication by tunable parameters. Our work is a proof-of-concept for this idea. While being only a measure for relevancy, we think that our contribution is useful for future Link Prediction methods with strict memory and computational requirements.

Our behavior-based relevance estimation model is evaluated over several artificial data sets and compared to a simple exponential decay function. Our evaluation's focus was to understand the behavior of our model in presence of obvious but slightly randomized mathematical distribution patterns.

The rest of this paper is organized as follows: Sect. 2 presents an overview of related work and briefly mentions approaches that incorporate time-based features and references to statistical process control methods that inspired our work. Section 3 will explain our model. Section 4 will demonstrate the results of our approach and provide advice for parameter tuning. Section 5 will finish our paper with a conclusion and an outlook for further work.

2   Related Work

A common example of estimating the relevancy of an edge between two nodes is the problem of *link prediction* [3]. Being applied to social network graph structures, where persons are represented as vertexes and their corresponding associations encoded by edges, link prediction tries to solve two classes of problems: First, whether two nodes that are not connected with each other yet *will* create an edge in the future (i.e., will become friends) or not. Second, whether two nodes will interact with each other *again*. This means that they already have a connection with each other and each edge encodes some kind of interaction (i.e., e-mails, phone calls, text messages, or co-authorship in research papers) [5]. The first problem can be expressed by a simple graph where a new edge is formed only between nodes with a distance of two or higher. For social networks, this can be understood as not yet connected persons who create a new link between each other. Regarding the second problem, new edges can also appear between nodes that already have a common edge to each other (sometimes referred to as repeated edges or repeated links) [6].

Focusing on networks that evolve over time, O'Madadhain et al. proposed a link prediction algorithm for event-based network data that honors temporal features [5]. Doing so, they demonstrated that the hierarchical structure of an organization as well as the likelihood of "co-participation" between persons can be inferred by interactions modeled as discrete events that contain sequential information and time data.

Further experiments published in [6] built upon that work and compared statistical approaches without time-based features with time-based counterparts. Their research yielded that even simple time-incorporating methods can outperform advanced statistical approaches on repeated link prediction.

Lankesh et al. [4] motivated that different forms of relationships have different kinds of decay—modeled as a power function. A long-lasting relationship formed by several discrete events over time might have a slower decay than a one-time contact.

Anomaly and outbreak detection [1] are well-known methods to discover emerging trends on Twitter, problems in system infrastructure, and behavioral change in response to changed circumstances.[1] We want to find a way to identify vertexes that interact with each other in a special way—that is, unexpected drops or raises in communication intensity. Looking for approaches, we found the domain of statistical process control being a valuable source for ideas.

We were originally experimenting with a model derived from the idea of the Western Electric Rules—an outlier detection approach where one assumes a normal distributed set of points. Each point's distance to the mean is measured in standard deviations. In a very brief summary, this approach is looking for patterns of consecutive points whose probability of occurrence is so low that the chance of being produced by the normal distributed process under investigation can be considered an anomaly. We realized that these rules cover only aggressive changes in the intensity of an event stream. However we have been inspired by the simplicity of their approach, as they combine simple statistical measures with interpretable rules.

---

[1]https://blog.twitter.com/2015/introducing-practical-and-robust-anomaly-detection-in-a-time-series.

3 Approach

The behavior-based relevance estimation model (*BBRE*) introduced in the following sections measures the relevancy of an edge over time. As new events of interaction come in (or be hollowed), an edge will be classified as (not) relevant. The final decision depends on a set of user-defined rules. Since relevancy varies on a per problem basis, we are going to define "relevancy" here at first, describing BBRE afterward.

We measure behavior by the intensity of interaction events (i.e., event count). An edge is considered relevant, if its behavior at a given point in time with respect to its recent behavioral history is of interest for a given problem under investigation. In order to use BBRE, one has to define whether a steady, falling, or rising intensity in behavior—or the transition from one of these states to another state—is of interest to the problem domain. Let's say two persons are represented by two different nodes in a social graph and exchanged messages between both persons denote that an edge is present between both nodes. In addition to that, one wants to find out whether these two nodes are developing positively. *In this example*, a positive development is defined as rising interest that possibly stabilizes around some level of interest. Falling interest that possibly stabilizes is not of interest (edges in this state are referred to as not relevant—again, only in this example). In addition to that, all these statements are relative to the recent interaction history—which means that their relevance changes according to their behavior over time.

BBRE has been optimized with respect to computational complexity and storage requirements with respect to the state that has to be managed by the model. One of our most essential biases for the proposed model is that the relevance of a relationship—with respect to the object of investigation—can be inferred by the development in intensity of the event stream which forms the relationship.

Whatever is considered a relevant behavior with respect to the problem domain in question, we will refer to edges that are deemed to be relevant as *active* as long as they *are* classified as relevant by our model and otherwise coin them *inactive*.

Different authors already demonstrated a quality-increasing effect of reflecting recency and other time-based features in their models [5, 6]. Therefore we'd like to introduce an additional effect to our model—a *social decay*.

3.1 The Model

We have to keep track of event occurrences, compute a sliding windowed mean, and the geometric mean over the ratio of change in the mean.

A *sliding windowed mean* is also referred to as *moving average*, *rolling average*, or *running average* in literature. It works by choosing a window width w (a size of a subset), adding elements ordered by time until the window is full, computing the average for the given subset, and then moving forward. Moving forward means that one drops the oldest element in order to make place for the newest—and computes the average again. This mitigates the impact of short-time outliers and highlights long-term trends. The same applies to a sliding windowed geometric mean over the ratio of change in the mean.

Let n be the total number of events and k the number of *remembered* events (more on that later) occurred *within* our given time window, then the rolling mean over k occurrences is defined as

$$rollingMean(n) = \frac{1}{k} \left(\sum_{i=n-k}^{n} event_i \right) \tag{1}$$

which is then used to calculate the rolling geometric mean over the *ratio of change in the rolling mean*:

$$geoMeanOfChangeInMean(n) = \left(\prod_{i=n-k+1}^{n} \frac{rollingMean(i)}{rollingMean(i-1)} \right)^{\frac{1}{k}} \tag{2}$$

BBREs ability to remember past events is limited in two ways: First by the width w of the window over time. This makes sure that events that occurred long time ago have a very low up to no influence on the current result. An additional feature of this technique is its robustness against short-time outliers provided that w is large enough. The window is divided into w slots—each has the size of one time unit. The second limit comes by the introduction of k—we are only able to fill up to k slots in our window w. Within each slot we record the number of events that have occurred within that slot.

We choose to use a ring buffer of size k where k denotes the maximum number of slots we can use to record event occurrences. In case the buffer is full but the oldest recorded event still falls into the time range denoted by w, the oldest event will be pruned to make place for the newest.

The following provides a brief description regarding the computational complexity:

1. Finding the oldest tuple that fits into the time window takes $O(logn)$.
2. Marking older elements in the buffer as stale takes $O(1)$
3. The structure requires a heap of $O(k)$ (+ a stale-marker pointing to an index— $O(1)$).
4. Computing rolling mean and rolling geometric mean take $O(s)$ where s is the number of events that can be considered as stale.
5. Inserting costs $O(logn)$ instead of $O(1)$, since an insert will trigger operations described in 1., 2., and 4.

This structure stores a tuple (point in time, aggregated event counts) and maintains a sliding windowed mean and its *geometric mean* over the *ratio of change in the mean* per sliding window slot. The latter key figures may need to maintain a ring buffer structure on their own.

This results in a data structure that remembers up to k points in time where one or more events occurred within a time window of a fixed, given length.

In addition to the outlined computational complexity properties and storage requirements, the choice of a ring buffer requires only initial memory allocation operations.

Provided that the parameters k and w are fix for an instance of a graph clustering algorithm, this structure's storage requirements will scale linear with respect to the number of edges in the graph as well as linear with respect to the number of processed events in the graph.

Social Decay, as used throughout this paper, refers to mitigating or eliminating the effect of very old events for the current relevancy estimation—enabling the model to forget. The approach of using a sliding window with limited memory as described in the section before implements this idea. Key figures are updated as events come in and are used to determine the relevancy of the edge. While we have not yet explained *how* to estimate an edge's relevancy (we are going to do this in the next section), we have a problem with this approach: If we only trigger this estimation as new events arrive—the last classification will persist as soon as no new events come in. Due to this, we introduce the following *decaying function* in order to define a maximum time an edge is considered *active* without further evidence (in the form of new events). Let t be the time since the occurrence of the n-th event.

$$geoMeanOfChangeInMean(n) \cdot e^{-\lambda \cdot t} \qquad (3)$$

Depending on the problem domain, we can restrict our classification by a maximum lifetime, a static threshold, or a dynamic threshold. Note: if the used threshold is relative to $geoMeanOfChangeInMean(n)$, Eq. (3) can be solved for t—which is a static value.

The parameter λ can be rewritten as $\tau = \frac{1}{\lambda}$ where τ represents a time frame in which $geoMeanOfChangeInMean(n) \cdot e^{-\lambda t}$ is reduced to a fraction of $\frac{1}{e}$ (approx. 37%). In physics this is referred to as *mean lifetime* and may help in finding a good problem specific value for λ.

3.2 Classification

An important property of the geometric mean is that it is agnostic to the scale of the observed values—it does not matter whether it deals with thousands, hundreds, or just a few events per time frame. Furthermore, it gives a ratio of change over the observed time window—we use this ratio of change in order to classify the *intensity* of a stream of events as falling, rising, or stable.

We introduce a threshold θ that defines the width of the margin for outcomes of the $geoMeanOfChangeInMean(n)$ that are considered stable. If you want to consider every development as stable that increases/decreases by 1%, θ would be 2%.

Every time a tuple (point in time, event count) is added to our structure, we compute $geoMeanOfChangeInMean(n)$ and classify its development into one of the following three trend states:

1. Falling, if the geometric mean falls below $1 - \frac{\theta}{2}$
2. Stable, if the geometric mean falls within $[1 - \frac{\theta}{2}, 1 + \frac{\theta}{2}]$
3. Rising, if the geometric mean is above $1 + \frac{\theta}{2}$

Given this classification, we can define activation and deactivation rules for an edge: We can (de-)activate an edge based on reaching/having/losing a certain state or transitioning from one specific (or unspecific) state to another.

During our work we achieved best results when we classified the activity state of an edge as described. In addition to that we always predicted a point in the future by utilizing Eq. (3), where the activity would toggle from *active* to *inactive* in case no new events would occur (we call these points *pivots*). In case a new event occurs that causes our classifier to re-label the edge as *inactive*, we checked whether the last *pivot* would have predicted this edge as *active* in case the given point would have not occurred. If so, we ignore the prediction of our model and still label the given point as *active*.

4 Experiments

We evaluated our approach over several artificial data sets of 10,000 data points. Points in time are distributed over an interval from zero to one to eliminate a bias based on a real-world time unit. A time unit of our experiment has the size of $\frac{1}{1000}$ of the whole time range of the experiment. BBRE is configured with $w = 0.06$ (a window of 6% of the observation time), $k = 30$, and $\lambda = 0.1$. Activation rules are triggered on raises as well as raises followed by stabilizations (but not on downtrends followed by stabilizations). We demonstrate three different activation thresholds with $\theta \in \{0.004, 0.005, 0.006\}$. We decided that our BBRE social decay fallback from Eq. (3) should drop as soon as it reaches 5%. Our choice of λ and this relative threshold reflect that we want to fade out the last impression of an interaction that paused or stopped to last at least 30 time units (see Sect. 3).

We compare our results to a simple classifier which utilizes an *exponential decay* function that increases its value at each event observation by one. If this function value is larger than a configured threshold $\theta_{compare}$, the edge is classified *active*. We set $\theta_{compare} = 100$ (1% of all data points) and $\lambda_{compare} \in \{50, 75, 100\}$.

As can be observed in Figs. 1a, b and 3b, the emphasized interaction development of our experiment has been well covered by the chosen parameter values while the baseline method $decay(events)$ performs with respect to a fine-tuned version of its corresponding $\theta_{compare}$ threshold better or worse. As can be seen in Fig. 1c, our configuration for BBRE 0.5% fails on understanding that the whole first peak should denote an outstanding activity compared to the overall history of the event stream. This can be explained in two ways: First the rolling window of width w has

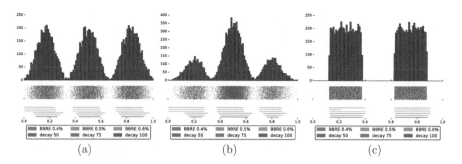

Fig. 1 (**a**) An event stream following a pattern generated by $\sin(x) + 1.1$ with three consecutive peaks. (**b**) A similar event stream with an outstanding second peak. (**c**) An event stream with two peaks surrounded and separated by noise. The colored lines below the chart denote when the graph is considered active by the corresponding classifier

been unable to cover that the block shows a more or less stable behavior—it gets distracted by local minimas and smaller deviations in interaction which is why it is unable to reactivate itself after a short downfall. This is due to the fact that the rolling window does not store enough observations and can be easily tricked by very short-lived peaks in behavior. A counter measure could be an adjustment of θ for larger deviations, a larger memory (increase in k and w) or a special treatment for trend evaluations in sparse windows, that is, using an outlier detection to identify single slots within w that influence the rolling mean in unwanted intensity. Second, our activation rules permit a reactivation. After deeper investigation of why BBRE 0.5% is the only configuration that fails at recognizing the whole first peak, we concluded that w is large enough but θ not sensible enough to recognize a strong enough *raise* in trend to reactivate itself. In fact, our 0.4% configuration classifies one bin as *raise* that has been classified as *stable* by the 0.5% configuration. On the other hand, the 0.6% configuration classifies another bin as *stable* that has been classified as *falling* by the 0.5% configuration. Since our activation rule is *activate on raise and on raise followed by stable bins, deactivate on falling and keep deactivated on falling and re-stabilization*, the behavior is expected though surprising on the first impression. This emphasizes the importance of bin size (a larger bin size avoids these edge cases) and the choice of the threshold.

The standard decay function has been demonstrated with three configurations. All of these configurations utilize thresholds that refer to absolute numbers in activity. While we have been working with data sets of static size (10,000 data points), a real-world use case would suffer from an unknown total size of an interaction stream. Figure 2 shows that *decay 75* is unable to cover the low-intensity middle block while *BBRE 0.4%* follows the behavioral change even *within* that block. All configurations of BBRE in the given example get distracted by a w that seems a bit too wide—since at the beginning of the second block, the window still contains data points of the first block. Rolling the window forward in time moves it to an area of *stable* intensity and keeps the edge deactivated until the first recognizable *raise*.

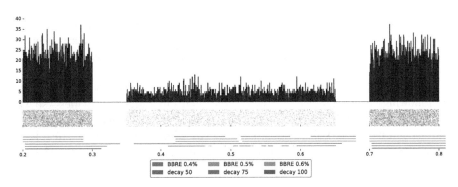

Fig. 2 A version of Fig. 3a with smaller bin size

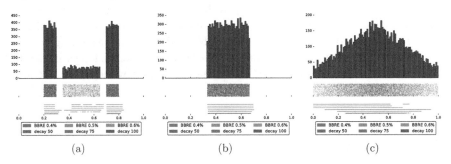

Fig. 3 (**a**) An event stream with three activity peaks surrounded by noise. (**b**) An event stream with one peak surrounded by noise. (**c**) An event stream with increasing activity approaching a momentum before falling down. The colored lines below the chart denote when the graph is considered active by the corresponding classifier

Let's assume Fig. 1a refers to a message activity stream from Monday to Wednesday during the working hours, a w wide and a k large enough to cover the range from 0.0 to 0.3 could be robust enough to understand that the given pattern is of recurring nature (the geometric mean of the rolling mean being near to 1.0) and therefore shows *stable* interactions. On the other hand, the same configuration applied to the data of Fig. 1b would signal an outstanding increase of interaction (the edge becomes active) while approaching the second peak. Arriving between 0.55 and 0.60, the data stabilizes after a very short but intense downtrend (since it arrives on nearly the same level as expected for stable behavior). From there on, a large w and k would treat the stream as stable again.

Figure 1a also demonstrates a conceptual difference between BBRE and the compared decay function configurations—while BBRE triggers a deactivation as soon as the downtrend gets strong enough, the decay approach is not able to recognize the behavioral change and focuses only on quantitative levels of interaction. The same can be observed in Fig. 3c.

One important decision during the application of our approach is the choice of a time unit. The bigger the scale of the unit, the less fine-grained patterns may be detected (compare a communication history aggregated over minutes, hours, days, or years).

Regarding the choice of parameters for our approach, we would like to suggest to following generalizations: The method only cares for ratios of activities—so we expect the same results for twice or half as much data as long as the data sets follow the same (observable!) distribution patterns. The window size w and k affect how much historic information is available at the time of prediction and the size of patterns of interest—the bigger these parameters are chosen, the more robust the results are regarding very short-lived changes in behavior. We would like to share the highly subjective opinion that a last impression of a relationship with much interaction is different to a relationship with fast decreasing interaction strength. In other words, having an intense interaction that suddenly stops is different than a sliding window of low interaction strength. A bigger λ will result in more aggressive activation state toggling. On the other hand, the smaller λ gets, the longer it takes to toggle in case no events occur—this can be interpreted as how long it takes to lose the power of the last impression.

5 Conclusion

We presented an approach to estimate the relevancy of interaction-based edges in a graph that evolves over time using information only available at a given edge without any further knowledge about the neighborhood of two nodes. As shown, our method is optimized with respect to low computational overhead and memory requirements. Moreover, a set of problem domain-specific rules can be defined to utilize the information about behavioral changes in an event stream to customize what kinds of behaviors are of interest to a given subject of investigation.

We demonstrated our relevancy estimator on several data sets with different characteristics regarding peaks, interaction pauses, and periodicity with success and compared our results with several configurations of a decay function.

One challenge of the decay function that has been compared to our approach is the estimation of λ and the θ_{decay} threshold. Though providing promising results when correctly chosen, different event streams need different parameter values in order to perform best for a given relationship, otherwise it fails on recognizing behavioral changes. Our relevancy method, on the other hand, shows more robust behavior for different data sets while having a little delay caused by the rolling window calculations. We think that future work should investigate whether it is possible to utilize our method in order to estimate a temporary λ and θ_{decay} for a given pair of nodes in order to combine the fast reaction time of the decay method with the robustness of our own approach.

References

1. Chandola, V., Banerjee, A., Kumar, V.: Anomaly detection: a survey. ACM Comput. Surv. **41**(3), 15:1–15:58 (2009)
2. Held, P., Kruse, R.: Detecting overlapping community hierarchies in dynamic graphs. In: 2016 IEEE/ACM International Conference on Advances in Social Networks Analysis and Mining (ASONAM), pp. 1063–1070 (2016)
3. Liben-Nowell, D., Kleinberg, J.: The link-prediction problem for social networks. J. Assoc. Inf. Sci. Technol. **58**(7), 1019–1031 (2007)
4. Munasinghe, L., Ichise, R.: Time aware index for link prediction in social networks. In: Cuzzocrea, A., Dayal, U. (eds.) Data Warehousing and Knowledge Discovery, DaWaK 2011. Lecture Notes in Computer Science, vol. 6862, pp. 342–353. Springer, Berlin (2011)
5. O'Madadhain, J., Hutchins, J., Smyth, P.: Prediction and ranking algorithms for event-based network data. ACM SIGKDD Explor. Newsl. **7**(2), 23–30 (2005)
6. Tylenda, T., Angelova, R., Bedathur, S.: Towards time-aware link prediction in evolving social networks. In: Proceedings of the 3rd Workshop on Social Network Mining and Analysis, p. 9. ACM, New York (2009)

Part II
Knowledge and Information Diffusion

Network Patterns of Direct and Indirect Reciprocity in edX MOOC Forums

Oleksandra Poquet and Shane Dawson

Abstract This paper proposes a set of indicators to capture the social context between regular forum posters in Massive Open Online Courses (MOOCs). Such indicators advance learning analytics research of social relations in online settings, as they enable to compare interactions in different courses. Proposed indicators were derived through exponential random graph modelling (ERGM) to the networks of regular posters in four MOOC forums. Modelling demonstrated that social context can be captured by the network patterns of reciprocity. In some MOOCs, network formation is driven by the higher propensity for direct reciprocity and lower propensity for indirect and triadic-level reciprocity. The social context in a highly moderated course was described by lower propensity for direct reciprocity and higher propensity for indirect and triadic-level reciprocity. We conclude that patterns of direct, indirect, and triadic-level reciprocity reflect variations of behaviour in knowledge exchange on MOOC forums. These three types of patterns can be theorised as dyadic information exchange, social solidarity, and gradual amplification of information flow. Re-modelling the same four MOOC networks without the staff and teaching assistants suggested that these network actors play a role in the formation of indirect and triadic-level reciprocity patterns, related to group cohesion.

Keywords MOOC forums · Interpersonal interactions · Reciprocity · Exponential random graph models

O. Poquet (✉)
Institute for Application of Learning Science and Educational Technology, National University of Singapore, Singapore, Singapore
e-mail: sasha.poquet@nus.edu.sg

S. Dawson
Teaching Innovation Unity, University of South Australia, Adelaide, SA, Australia

© Springer International Publishing AG, part of Springer Nature 2018
R. Alhajj et al. (eds.), *Network Intelligence Meets User Centered Social Media Networks*, Lecture Notes in Social Networks,
https://doi.org/10.1007/978-3-319-90312-5_4

1   Introduction

Massive open online courses (MOOCs) are educational provisions that enable access to professional development and lifelong learning for millions of individuals across the globe. Instead of formal credentials, MOOCs provide free or low-cost structured opportunities to learn for self-fulfillment or upskilling. Such unaccredited learning opportunities are intentional from the learner perspective, and structured through the learning objectives, course time and support. The context and characteristics of MOOCs are similar to other non-formal educational offerings, that is non-accredited courses whereby "learners opt to acquire further knowledge or skill by studying voluntarily with a teacher who assists their self-determined interests, by using an organised curriculum, as is the case in many adult education courses and workshops" [1].

The non-formal nature of MOOCs influences the social context generated through forum participation. By "social context," we refer to interpersonal peer-to-peer and peer-to-instructor interactions characterised by a relational quality such as emotional attachment. By operationalising interpersonal interactions as "social context," we emphasise that they are a part of the group communication history and include personal understanding of this communication. In other words, a social context is comprised from interpersonal interactions but qualitatively reaches beyond simple information exchange.

In the context of formal education, where a bounded group of individuals interact continuously from day 1 of the course, interpersonal interactions and the emergent social context are tightly coupled. That is, the set-up of a formal course creates the necessary structure and boundaries for learners and teachers to interact in discussion forums, or obligatory assignments and tutorials. As a result, for formal education offerings, interpersonal interactions can readily evolve beyond simple information exchanges to shared understanding and history to enable social support, deeper learning, and motivation at later stages. In MOOCs, however, such emergent quality of interpersonal interactions cannot be assumed alongside aggregated patterns of interpersonal interactions. In MOOCs, the group boundaries are ill-defined, and participation in social interactions is intermittent at best. Given these differences, not all interpersonal interactions contribute to the quality of the emergent social context. Furthermore, analysis of the social context in non-formal educational settings needs to reflect the peculiarities of the MOOC context, rather than replicate approaches previously applied in formal education.

Current study examined the social context in four edX MOOC forums. Instead of including all forum interactions, analysis was limited to the interactions between individuals who regularly posted on the forums as the course progressed. The networks of regular MOOC posters were modelled through exponential random graph modelling (ERGM) using the network patterns of direct, indirect and triadic-level reciprocity. Results suggested that reciprocity patterns differentiated the interaction structures in the four analysed courses. The study also highlighted reciprocity patterns associated with the teaching staff that allows hypothesising their role in the network formation. Suggested network indicators of reciprocity can be used to describe MOOC forums and compare networks of aggregated interactions between the learners by the end of the course.

2 Structure of MOOC Networks

This section explains how reciprocity network patterns can model the structure of MOOC regular posters. We review the literature on (1) conceptualization of informal knowledge-exchange online communities; (2) reciprocity, with the focus on indirect reciprocity as reflecting altruistic motivation; and (3) network structures in both Electronic Networks of Practice (ENP) and MOOC forums. Through the literature review, we argue that patterns of direct reciprocity (knowledge exchange), indirect reciprocity (altruistic contributions) and triadic-level reciprocity (amplification of network flow) can describe different structures of MOOC forum networks.

2.1 Electronic Networks of Practice (ENP)

The formation of informal online communities has been previously explained through individuals' altruistic motivation to contribute to the collective benefit [2]. Faraj, Wasko and Johnson examined the so-called electronic network of practice defining this as "a self-organizing, open activity system that focuses on a shared interest or practice and exists primarily through computer-mediated communication" [3]. Wasko and Faraj [2] conceptualised ENP as sustained by voluntary participation and knowledge exchange. Participants of these informal online communities shared knowledge about particular interests that comprised group purpose.

In addition to the attachment to the group's purpose and knowledge exchange around a shared practice, such communities have been associated with altruistic behaviour. That is, in online communities based on shared purpose or interest, people seek information by relying on the "kindness of strangers" [4]. Specifically, Constant et al. [4] found that the employees of a global computer manufacturer participating in an ENP gave useful advice and solutions to the problems of others despite a lack of personal connections. Some 81% of the providers did not know the seekers, and 10% were barely acquainted. Hence, there was no direct benefit derived for a poster providing a solution. Altruistic motivation can explain why individuals make such contributions. In effect, the solutions are offered for the sake of the community and a wider public good. Similarly, Faraj and Johnson [5] suggested that the information exchanges within an ENP have a social, that is group-oriented, motivation. The authors contended that interactions among the participants were more than just information queries, and provided a social purpose.

Their contributions were socially oriented towards a broader community benefit. Simply put, individuals participating in an ENP were aware that the larger group would benefit from their reply to an individual information-seeker. They provided the information in ways accessible to others. The underlying altruistic motives were also observed in a range of contexts demonstrating a group-oriented attitude among dominant motivations of ENP members.

2.2 Interpreting Network Patterns Through Reciprocity Construct

Previous work in network science allows representing altruistic behaviours as local patterns of a network. In particular, altruistic behaviour can be captured through the network forms of reciprocity. For instance, the altruistic "pay-it-forward" behavior, that is providing a public good without an expected return, is one of the underlying forms of reciprocity. Other reciprocity types differ in the extent of altruism, that is as to when an action is expected to be reciprocated, if at all. For instance, when both parties negotiate the terms of a bilateral exchange as in "if you scratch my back, I'll scratch yours", such reciprocity is considered "negotiated," and a return of the service is expected [6]. According to the collective action theory, such an exchange indicates a non-altruistic motivation of self-interest [7]. When the direct reciprocation of one's action could be delayed, or not offered at all, Molm [8] described such a form of exchange as unilateral. Unilateral reciprocity implies that although the possibility that the other person may reciprocate exists, there is a level of uncertainty as to when or if this may occur. Indirect reciprocity, a type of unilateral reciprocity also referred to as generalised exchange [9], is represented by the network form where individual "A may give to B on one occasion and to C on a different occasion". Indirect reciprocity can also be represented by the following patterns "A gives to B, B gives to C" as well as "A gives to B, B gives to C, and C gives to A" [6]. Examples of such forms of collective exchange can be found, for instance, in voluntary blood-giving activity or academic peer review practices.

The network configurations associated with indirect reciprocity are more than just structural representations. The forms of reciprocity carry meaning about the socio-emotional quality of the exchange. One such quality is the underpinning trust that characterises the forms of unilateral reciprocity. In a unilateral exchange when a service is reciprocated directly or indirectly, the actor provides material or symbolic goods without knowing "whether, when, or to what extent the other will reciprocate" [6, 8, 10]. Since an individual is unsure about the return of the service, an increased possibility of risk is another characteristic of a unilateral exchange (direct and indirect reciprocation). Due to such embedded risk, such reciprocal forms are also associated with a higher emotional value. By reciprocating without an expectation of return, individuals demonstrate their readiness to pay forward a service and show a willingness to continue the relationship. According to Molm [8], the forms of a unilateral exchange (direct and indirect reciprocity) take place in settings where conflict is of low salience. That is, if conflict is minimal, individuals are more likely to offer services unilaterally. Molm et al. [6] further validated that "paying it forward" to other group members, represented by the indirect reciprocity forms, occurred in contexts with higher levels of social solidarity, than those described by direct or negotiated reciprocity.

2.3 Network Patterns of Reciprocity in ENP and MOOCs

Prior research identified the presence of reciprocity forms in network structures of ENPs and all-poster MOOC networks. Both direct and indirect reciprocity configurations were found pertinent to informal online identity-based communities. Specifically, in modelling the ENP, Wasko et al. [11] described their structure through the dominant features of direct reciprocity over indirect reciprocity. Similarly, the features of direct reciprocity were found prominent in describing the structure of all-poster MOOC networks [11–13]. Joksimovic et al. [14] suggested that besides network forms associated with direct and indirect reciprocity, the all-poster networks can be defined by a simmelian tie form. We suggest that this form is interpreted as triadic-level reciprocity, that is the mutual exchange of interactions at the triadic level (A → B, B → A, B → C, C → B, C → A, A → C), and represents a clique-like clustering in a network, and its gradual amplification.

2.4 Research Questions

Due to the prominence of the features of direct and indirect reciprocity observed in informal online identity-based communities, the study examined the role of reciprocity features in non-formal educational online communities for the network formation through the following research questions:

1. How do network patterns of reciprocity describe the social structure of emergent networks of regular posters in MOOCs?
2. Were course staff, instructor, and assistants embedded in a particular network representation of reciprocity?

3 Methods

To address research questions, we have applied ERGM to four networks of regular posters in MOOCs. To address the second research question, we have removed teaching staff and assistants from the networks and applied ERGMs to the new networks. It is evident that removing the teaching staff does not exclude the effect of their presence on the other ties. As such our interest was to identify the network structures where the teaching staff were situated. This section elaborates on the network construction and the features of ERGMs.

3.1 Network Construction

Defining the Edges

Network ties were non-valued and directed. The direction of the social tie represented the collective orientation of communication flow. That is, posts to the forum were considered as a contribution offered to the collective group engaged in the discussion rather than a reply or discrete message to an individual poster. Consequently, network ties were directed to all the posters in a single thread who preceded the actor based on timestamps. If A posted, B replied, and C commented, and D replied, each of the subsequent actors would have a directed tie to everybody else prior to them. That is, B → A, C → B, C → A, D → C, D → B, D → A. Such representation of interactions through one-to-many ties reflects the principle of collective orientation where an individual contributed information to the posting group (or rather collective discourse within a particular thread), rather than provided a service to an identified individual (as replying to a known person is often not feasible in a MOOC discussion thread).

Defining the Nodes

The purpose of this study was to model the structure of the evolving communities in MOOC forums. Therefore, the inclusion of nodes into the network was guided by the potential of the learner (i.e. node) to develop interpersonal relationships that are typical for communities. As a result, not all forum posters were included as nodes in the constructed networks. In our previous work [15–17], we argued that forum interactions in MOOCs are qualitatively different from one another, depending on the regularity of forum posting activity by the learner contributing them. As elaborated in Poquet [15], an average of 2.9% of forum posters have potential to establish interpersonal relationships with one another exclusively through in-forum interactions. These learners establish a familiarity with others in the group due to their continuous/regular posting on the forums. In essence, as these individuals regularly post and receive more timely replies, there is increased opportunity to co-occur with one another over the course duration more so than more irregular and intermittent posters. Therefore, these "regularly posting" learners experience interactions that are effectively embedded within a shared history (in contrast to other posters) that emerges through wide-ranging topics of discussion covering content- and non-content-related issues [15]. We posit that the non-valued edges between the regular posters can be defined by such parameters, whereas the edges between the other posters may not. We refer to this group as the forum core (FC), and assume that due to the quality of the interactions the individuals constituting FC may evolve into a community.

Temporal Boundaries

Network boundaries were set around the course delivery times: between the week the first video lecture was released, and the week when the last lecture in the course was provided. As communication between actors can potentially continue well past a course's lectures due to exams or assignments, any established closing date for the forum analysis would have been arbitrary. The discrete temporal limits applied in this study made the comparison of courses more feasible.

3.2 Exponential Random Graph Modelling

The network properties describing the structure of forum cores' networks were used by applying exponential random graph modelling (ERGM). ERGM is a probability model (p*) for network analysis [18] that predicts the presence of the ties and local structures in networks. ERGM estimates the probability of a network G, given the sum of network statistics weighted by parameters inside an exponential [18]. In ERGMs, network structure is understood as emergent from (1) *endogenous structural parameters* associated with social processes, such as, for example, preferential attachment or reciprocity, and (2) *exogeneous parameters*, such as individual actor attributes or the impact of interpersonal relations in another network. The network statistics used to reproduce *endogenous structural parameters* represent social processes expressed in sets of two actors (dyads), three actors (triads) or larger combinations, such as several triads sharing a tie (social circuits). An ERG model estimates the likelihood that each selected configuration, associated with a theoretically formulated social process, will occur beyond chance in a random network generated based on the modelled constraints.

3.3 Modelled Configurations and Effects

Structural Configurations

The study examined the propensity of three structural configurations to form the structure of forum core's networks, listed along with the correspondent statnet label [19]:

Table 1 Descriptive statistics of structural network configurations

Network statistic	Course name			
	Water, N	Solar, N	FP, N	Excel, N
Nodes, FC	340	254	161	220
Edges, FC	4574	3059	2978	2259
Direct reciprocity	725	629	880	516
Indirect reciprocity	3442	2626	1734	2084
Mutual exchange at triad level	1164	1060	2963	882
Nodes, FC w/out staff	340	247	157	213
Edges, FC w/out staff	4574	2061	2398	1203

Note: *FC* forum core

- Direct reciprocity (A → B, B → A; *mutual* label)
- Indirect reciprocity (A → B, B → C; *cyclicalties* label)
- Reciprocal exchange at the triad level (A → B, B → A, B → C, C → B, A → C, C → A; *simmelianties* label)

These network configurations structurally control for the network reciprocity, degree and closure. In our previous work [17], we have modelled MOOC forum core networks without the indirect reciprocity pattern. Although this study demonstrates somewhat similar results, we argue that including indirect reciprocity feature is essential to understanding why the network forms. Moreover, from a network point of view including an indirect reciprocity pattern is a must as it controls for degree distribution within the structural components of ERGM. Table 1 presents the counts of modelled network configurations for each of the networks.

Node Attributes

In addition to controlling for the structural parameters of reciprocity, the models also controlled for the effect of learner super-posting activity on the propensity to send and receive ties. The distinct levels of activity across online networks have been previously supported in empirical studies. Studies of Internet communities captured the individuals who represent a large proportion of the overall activity [20]. Similarly, in MOOCs, several studies have shown the presence of hyperposting activity [21, 22]. For this reason, modelling a forum core needed to account for a very active participation level of a select few, and a relatively low level for the overall majority. The *nodefactor* term was used for modelling the effect of participation level, that is no distinction was made as to whether the learner sent or received ties.

To identify the overall level of activity, instead of relying on the number of messages posted, we used measures of learner activity (1) as devised by Hecking et al. [23]. Hecking and colleagues established in-reach and out-reach metrics of individual activity occurring in a MOOC network [23, 24]. These measures calculate the diversity of outgoing and ingoing relations for a node i, where $w(e_{i,j})$ represent the weight of the tie between them, od(i) and id(i) represent out-degree and in-degree, respectively.

$$\text{outreach}(i) = \text{od}(i) - \sum_{j \in \text{outneigh}(i)} w\left(e_{i,j}\right) \times \log\left(\frac{w\left(e_{i,j}\right)}{od(i)}\right)$$
$$\text{inreach}(i) \;\; = \text{id}(i) - \sum_{j \in \text{inneigh}(i)} w\left(e_{i,j}\right) \times \log\left(\frac{w\left(e_{i,j}\right)}{id(i)}\right)$$
(1)

This study replicated this measure in order to identify the in- and out-reach of an individual's activity across all-posters in MOOC forums. By using k-means clustering of in-reach and out-reach measures per person, all forum posters were divided into three groups: to represent highest, moderate and low forum participation activity in the entire network. The posting activity attributes were used to control for tie propensity formation in forum core ERGMs. Table 2 presents the descriptive statistics of actors within each cluster (counts), along with their average message posting activity. Cluster 1 refers to learners with low posting activity. Cluster 2 refers to the posters with moderate posting activity, and Cluster 3 represents hyperposters.

Labelling of activity levels (Table 2) was relative to the course, and heterogeneous in measures across the courses. That is, a hyperposter in one course may have similar activity numerically as a moderate poster in another course. As shown in Table 2, the Water course had the lowest level of moderate and hyperposting activity (40 posts per person and 218 posts per person, respectively) and FP had rather high level of posting activity for low level and moderate level of participation within its network (20 posts per person, and 125 posts per person, respectively). However, on average, the proportion of activity across the three clusters within the courses appeared to be relatively comparable.

Goodness of Fit

Several steps to ensure goodness of fit were conducted [18, 25]. That is, converged networks with the best fit were checked for degeneracy. For this study's analysis, R packages statnet and ergm were employed [19, 26]. A Monte Carlo Markov Chain (MCMC) algorithm was used to check for network degeneracy and to examine the goodness of fit. Estimated models demonstrated reasonable goodness of fit (1) as the mean number of networked configurations was similar to configurations in the observed network; (2) the MLE estimation produced non-degenerate networks; and (3) the AIC coefficient demonstrated better fit than the null model (Table 3). Goodness of fit of the modelled network was lower in forum cores with the teaching assistant removed. This was particularly pronounced for the Excel course, where forum core network without staff was also lacking all the nodes defined by hyperposting activity.

Table 2 Descriptive statistics of the three clusters representing forum core posting activity levels in forums

	Cluster characteristics	Poster clusters		
		Cluster 1	Cluster 2	Cluster 3
Water	*Posters, N*			
	• All-poster network	2998	273	6
	• Forum core	238	96	6
	Posts per poster, M(SD)	15 (9)	40 (27)	218 (45)
Solar	*Posters, N*			
	• All-poster network	4082	40	7
	• FC	221	28	5
	• FC, w/out course staff	219	24	4
	Posts per poster, M(SD)	18 (14)	129 (95)	586 (238)
Excel	*Posters, N*			
	• All-poster network	1994	38	3
	• Forum core	201	16	3
	• FC, w/out course staff	200	13	0
	Posts per poster, M(SD)	13 (12)	115 (108)	534 (286)
FP	*Posters, N*			
	• All-poster network	1317	41	4
	• Forum core	130	27	4
	• FC, w/out course staff	128	26	3
	Posts per poster, M(SD)	20 (17)	125 (178)	459 (188)

Note: *FC* forum core

Cluster 1 represents low-level participation, cluster 2 moderate-level participation and cluster 3 high-level participation

Water does not have descriptive statistics of FC without teaching staff because the course was unmoderated and had no instructor/teacher assistants' activity

4 Results

The first research question inquired if patterns of direct reciprocity associated with information exchange and of indirect reciprocity associated with altruistic motivations can describe networks of regular posters in MOOCs. The networks were modelled through the baseline density of networks, direct reciprocity of ties, indirect reciprocity of ties, and reciprocal exchange at the triadic level. The models also controlled for the propensity of the formation of ties sent and/or received by the posters with different levels of network in-reach and out-reach, reflective of their overall posting activity. The levels were distinguished between individuals with low, moderate and high forum participation behaviour. All models were fitted using the described network configurations.

The estimates for the network parameters of modelled MOOC FC networks are outlined in Table 3. There was a significant and positive effect for direct reciprocity in all four FC networks, namely [Water: 1.27(0.009), Solar: 1.55(0.12),

Table 3 Outputs for final ERG models in four edX MOOCs

	Course name			
	Water	Solar	Excel	FP
Nodes, N	340	254	220	161
Edges, N	4574	3059	2260	2978
Structural effects				
	Estimate (SE)	*Estimate (SE)*	*Estimate (SE)*	*Estimate (SE)*
Density	−4.34*** (0.03)	−4.38*** (0.05)	−4.69*** (0.07)	−4.76*** (0.24)
Mutual reciprocity	1.27*** (0.09)	1.55*** (0.12)	2*** (0.15)	1.02*** (0.28)
Indirect reciprocity	0.2*** (0.03)	0.7*** (0.03)	0.81*** (0.06)	1.54*** (0.24)
Triad level reciprocity	0.53*** (0.04)	0.54*** (0.05)	0.22*** (0.05)	0.83*** (0.13)
Effects of node participation level (low-level activity as intercept)				
Moderate posting	0.77*** (0.02)	0.68*** (0.02)	1.17*** (0.04)	0.73*** (0.02)
Hyperposting	1.5*** (0.05)	0.87*** (0.04)	1.7*** (0.07)	1.17*** (0.05)
AIC null	38,485	24,601	18,244	18,450
AIC final	32,875	20,262	13,690	14,308
BIC final	32,933	20,317	13,743	14,357

Note: Each course network was modelled separately, only the final models are presented
** Stands for p-value of <0.001

Excel: 2(0.15), FP: 1.02(0.28)], with estimates reported for each course and the standard errors in parentheses. There was a significant and positive effect for indirect reciprocity in all four forum core networks [Water: 0.2 (0.03), Solar: 0.7(0.003), Excel: 0.81(0.006), FP: 1.54(0.24)]. There was also a positive and significant effect for reciprocal exchange at triad level [Water: 0.53 (0.004), Solar: 0.54(0.05), Excel: 0.22(0.05), FP: 0.83(0.13)]. All the estimates for the four models were significant, with a p-value of <0.001.

The results supported the premise that features of reciprocity can reflect the structure of forum core networks, much like in online public electronic groups. The forum cores' networks in the three analysed courses were more defined by a higher propensity for direct dyadic exchange [Water: 1.27(0.009), Solar: 1.55(0.12), Excel: 2(0.15)] in contrast to a lower propensity over tie formation of indirect exchanges [Water: 0.2 (0.03), Solar: 0.7(0.003), Excel: 0.81(0.006)] or triadic-level exchanges [Water: 0.53 (0.004), Solar: 0.54(0.05), Excel: 0.22(0.05)]. In sum, this reflects the expected network structure observed in informal online communities and all-poster MOOC networks. Such findings generally support the presence of hypothesised social processes of reciprocity in a forum core. In general terms, a lower indirect reciprocity indicates a lower social solidarity, and potentially a lower level of maturity for an identity-based community within a forum core.

However, for the FP course such dynamics were not replicated. The FP course had a higher propensity of indirect exchanges [1.54(0.24)] to form ties than a direct reciprocity configuration [1.02 (0.28)]. Also, FP's estimate for mutual exchanges at a triadic level was higher than in other courses [Water: 0.53 (0.004), Solar: 0.54(0.05), Excel: 0.22(0.05), FP: 0.83(0.13)]. FP's higher propensity for network

Table 4 ERGM outputs for the remodelled three forum cores, where staff and teaching assistants were removed from the network

	Solar	Excel	FP
	Estimate (SE)	Estimate (SE)	Estimate (SE)
Density	−4.26*** (0.04)	−4.43*** (0.04)	−3.81*** (0.08)
Mutual reciprocity	2.09*** (0.11)	2.46*** (0.12)	1.37*** (0.18)
Indirect reciprocity	0.52*** (0.04)	0.6*** (0.04)	0.57*** (0.07)
Triad-level reciprocity	0.36*** (0.05)	0.29*** (0.06)	0.63*** (0.08)
Moderate posting	0.49*** (0.03)	0.66*** (0.05)	0.62*** (0.02)
Hyperposting	0.79*** (0.05)	NA	1.02*** (0.05)
AIC null	18,001	11,098	15,700
AIC final	15,943	9778	12,623
BIC final	15,997	9822	12,672

Note: Each course network was modelled separately, only the final models are presented
NA network contained no nodes of hyperposting activity, and they were not modelled
**Stands for *p*-value of <0.001

closure can be interpreted as a more evolved network development and FP's higher level of indirect reciprocity can be interpreted as a network form of social solidarity. The propensity for these two levels of reciprocity to describe the forum core's structure in the FP MOOC suggests that this social group had a different social dynamics underpinning its network formation.

The second research question inquired into the types of network structures that embedded the teaching staff. Three networks that originally included the staff and teaching assistants were removed, and the networks were re-modelled to highlight the role of the instructors in observed network structures. All additional models also converged on the modelled parameters and were not degenerate. The estimates for the network parameters of modelled MOOC FC networks where staff and teaching assistants were removed are outlined in Table 4. There was a significant and positive effect for direct reciprocity in all four FC networks, namely [Solar: 2.09(0.11), Excel: 2.46(0.12), FP: 1.37(0.18)], with estimates reported for each course and the standard errors in parentheses. There was a significant and positive effect for indirect reciprocity in all four forum core networks [Solar: 0.52(0.004), Excel: 0.6(0.004), FP: 0.57(0.07)]. There was also a positive and significant effect for reciprocal exchange at triad level [Solar: 0.36(0.05), Excel: 0.29(0.06), FP: 0.63(0.08)]. All the estimates in four models were very significant, with a *p*-value of <0.001. Water FC was not modelled as the network contained no staff teaching assistants that interacted with students regularly, and the forums were unmoderated.

To aid interpretation, Fig. 1 plots the odds of structural patterns occurring in network structures with teaching staff and without. Figure 1 demonstrates that the teaching staff are largely embedded within indirect reciprocity and triadic-level reciprocity structures. Such results may suggest that the instructor and teaching assistants play a role in promoting altruistic behaviour and clustering of the network leading up to the more evolved social context. Yet, further analyses more fine-tuned to investigating the role of teaching staff are required to draw further conclusions.

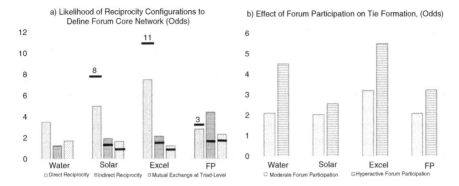

Fig. 1 Cross-course comparison of the social processes driving the formation of a forum core (FC). The black solid line indicates odds of the network configurations in FC if the staff and teaching assistants were removed

5 Discussion

The study examined the structure of four MOOC networks of regular posters using the patterns of reciprocity as indicators of the emergent social context in MOOC forums. We argued that the network configurations can capture the different behaviors that comprise social context in MOOC forums. Specifically, we argued that MOOC networks of regular learners could be described through direct reciprocity to capture dyadic knowledge exchange, indirect reciprocity to capture altruistic behaviour, and network closure comprised of direct and indirect reciprocity. The first research question demonstrated that these reciprocity patterns could describe the structure of MOOC networks of regular posters. Three analysed networks were defined by a higher propensity for direct dyadic exchange, in contrast to a lower propensity over tie formation of indirect exchanges, or triadic-level exchanges. This reflects the expected network structure observed in informal online communities and all-poster MOOC networks. Such dynamics were not replicated in one network that was defined by a higher propensity for network closure and higher level of indirect reciprocity. This network's structure is indicative of a more evolved social context, that is defined by a high social solidarity indicator, represented by indirect reciprocity configurations, and a high group cohesion indicator, represented by triadic-level closure.

The second research question inquired about the structural patterns that embedded teaching staff. Analysis demonstrated that the teaching staff was embedded within the indirect reciprocity structures and triadic-level reciprocity. However, current analysis did not address the effect that the presence of instructors had on the formation of ties among learners. Further work is needed to determine the extent to which teaching staff is instrumental to the network's clustering and subsequent amplification.

The social structures observed in FP can be described as the course with the highest level of social context development among the four analysed MOOCs, given the following interpretation of the forms of indirect reciprocity and reciprocal closure. Molm [8] proposed three conditions that describe environments with indirect reciprocity. First, the risk of vulnerability is heightened, that is individuals are vulnerable in offering a service that holds no guarantee of reciprocation. Second, the expressive value of giving is higher, as the act of "paying forward" demonstrates a commitment to the shared practice by contributing without expecting anything back. Third, indirect reciprocity takes place when a conflict is less salient, that is the social tensions do not prevent contributors to "pay forward." Reciprocal closure, or reciprocation at the triadic level, is more representative of group-like formations that are taking over a network. An amplification of information flow, less control over information within a selected few and a higher integration of participants are aspects associated with such a structure. In sum, the combination of higher indirect reciprocity and triadic reciprocity could be interpreted as social solidarity and group cohesion.

Analyses suggest that there may be a positive relationship between hyperposting activity and direct reciprocity, as well as a negative relationship between hyperposting activity and indirect reciprocity. For the former, the pattern of a high propensity of hyperposters to send and receive ties, high direct exchanges and low triad-level exchanges is observed in three of the four examined courses (Water, Solar and Excel). For the latter, two courses (Water and Excel) with a higher direct reciprocation and higher hyperposting impact on tie formation have low indirect reciprocation. Both relationships can be interpreted as the impact of hyperposters impeding group formation within the networks, despite hyperposters' effort to stimulate discussions, or help those seeking answers. Higher levels of indirect reciprocity would indicate that learners feel compelled to reciprocate a perceived "favour" they may have received. However, if hyperposters assume responsibility for actively taking on questions, the overall group may not feel compelled to reciprocate to unknown strangers, or develop a common identity to learn and help others learn. In other words, hyperposting individuals may be diminishing the opportunities associated with a "pay it forward" behaviour.

What seems to be a negative relationship between social solidarity and high hyperposting impact does not indicate that moderation is detrimental for an evolving relational quality of interpersonal interactions. In line with prior research and practice of group formation, the role of a teacher, a moderator or a leader is critical at early stages of group formation. In initial course stages, moderation (e.g. Excel, Solar, FP) could stimulate direct reciprocity, as well as indirect reciprocity. The trend of a higher propensity to reciprocate is observed across all moderated courses that were examined in this study, as opposed to the unmoderated one (i.e. Water). However, while a critical mass of moderate posters may be willing to pay effort and actions forward (in their posting activity), they still require opportunities to do so. If a moderator continues addressing all possible queries posted in a MOOC forum, for example in the Excel course, at an incredibly high propensity, affordances for network closure are limited. In other words, as much as moderation is needed at the onset of a course, a moderator's activity may need to be less involved further along a course's timeline, in order to allow for reciprocal closure.

Much of the interpretations and their associated implications remain speculative. Although we have highlighted that the network structures of the regular posters in MOOC forums can be defined by the features of reciprocity, the sample examined in the study is rather small. Further analyses are required to both validate the set of reciprocity indicators as telling of MOOC forum social context, as well as better understand the relationships hypothesised in the Discussion section. Moreover, the second research question suggested that teaching staff plays an instrumental role in the formation of the network, while some aspects of the initial analysis demonstrate how teaching staff may impede such formation. Understanding the turns within the network formation and the role of the staff within them, both require more advanced SNA analyses, as well as other methodologies complementing the insights.

6 Conclusions

The study examined if network patterns of reciprocity can define the social structure of MOOC forum posters. Findings support the premise that these network structures can be described by the various reciprocity patterns and, in turn, could be integrated as network-level analytics of evolving social context in MOOC forums. Indicators of direct reciprocity represent dyadic information exchange, indirect reciprocity represents social solidarity and reciprocal closure indicates amplification of a network and a shift to a more egalitarian structure from a centralised one. The study demonstrated that courses where a designated moderator was involved exhibited higher indirect reciprocity than direct reciprocity in network formation. The formation of a forum core network in a highly moderated course was also mostly defined by an amplification of information flow reciprocally, at the triadic level. The study also suggested that hyperposting activity may have prevented networks from clustering, and hence impede a more egalitarian information flow.

References

1. Commission of the European Communities: a memorandum on lifelong learning, http://www.voced.edu.au/content/ngv:2687, (2000)
2. Wasko, M.M., Faraj, S.: Why should I share? Examining social capital and knowledge contribution in electronic networks of practice. MIS Q. **29**, 35–57 (2005)
3. Faraj, S., McLure Was M., Johnson S.L.: Electronic Knowledge Networks: Processes and Structure. In I. Becerra-Fernandez, D.E. Leidner (eds.) Knowledge management: An evolutional view. Advances in Management Information System. **12**, 270–292. Armonk, NY (2008)
4. Constant, D., Sproull, L., Kiesler, S.: The kindness of strangers: the usefulness of electronic weak ties for technical advice. Organ. Sci. **7**, 119–135 (1996)
5. Faraj, S., Johnson, S.L.: Network exchange patterns in online communities. Organ. Sci. **22**, 1464–1480 (2011)
6. Molm, L.D., Collett, J.L., Schaefer, D.R.: Building solidarity through generalized exchange: a theory of reciprocity 1. Am. J. Sociol. **113**, 205–242 (2007)
7. Ekeh, P.P.: Social exchange theory: the two traditions. Heinemann, London (1974)

8. Molm, L.D.: Theoretical comparisons of forms of exchange. Sociol. Theory. **21**, 1–17 (2003)
9. Molm, L.D., Peterson, G., Takahashi, N.: Power in negotiated and reciprocal exchange. Am. Sociol. Rev. 876–890 (1999)
10. Molm, L.D.: The structure of reciprocity. Soc. Psychol. Q. **73**, 119–131 (2010)
11. Wasko, M.M., Teigland, R., Faraj, S.: The provision of online public goods: examining social structure in an electronic network of practice. Decis. Support. Syst. **47**, 254–265 (2009)
12. Joksimović, S., Dowell, N., Skrypnyk, O., Kovanović, V., Gašević, D., Dawson, S., Graesser, A.C.: How do you connect?: analysis of social capital accumulation in connectivist MOOCs. In: Proceedings of the Fifth International Conference on Learning Analytics and Knowledge, pp. 64–68. ACM, Poughkeepsie (2015)
13. Kellogg, S., Booth, S., Oliver, K.: A social network perspective on peer supported learning in MOOCs for educators. Int. Rev. Res. Open Distrib. Learn. **15**, 263–289 (2014)
14. Joksimović, S., Manataki, A., Gašević, D., Dawson, S., Kovanović, V., de Kereki, I.F.: Translating network position into performance: importance of centrality in different network configurations. In: Proceedings of the Sixth International Conference on Learning Analytics & Knowledge, pp. 314–323. ACM Press, New York (2016)
15. Poquet, O.: Social context in MOOCs (2017)
16. Poquet, O., Dawson, S.: Analysis of MOOC forum participation. In: Australasian Society for Computers in Learning in Tertiary Education Conference (Ascilite) (2015)
17. Poquet, O., Dawson, S., Dowell, N.: How effective is your facilitation?: group-level analytics of MOOC forums. In: Proceedings of the Seventh International Learning Analytics & Knowledge Conference, pp. 208–217. ACM, New York (2017)
18. Lusher, D., Koskinen, J., Robins, G.: Exponential random graph models for social networks: theory, methods, and applications. Cambridge University Press, Cambridge (2012)
19. Handcock, M.S., Hunter, D.R., Butts, C.T., Goodreau, S.M., Morris, M.: statnet: software tools for the representation, visualization, analysis and simulation of network data. J. Stat. Softw. **24**, 1548 (2008)
20. Turner, T.C., Smith, M.A., Fisher, D., Welser, H.T.: Picturing Usenet: mapping computer-mediated collective action. J. Comput. Mediat. Commun. **10**, JCMC1048 (2005)
21. Huang, J., Dasgupta, A., Ghosh, A., Manning, J., Sanders, M.: Superposter behavior in MOOC forums. Proc. First ACM Conf. Learn. Scale Conf. **14**, 117–126 (2014)
22. Skrypnyk, O., Joksimovic, S., Kovanovic, V., Gasevic, D., Dawson, S.: Roles of course facilitators, learners, and technology in the flow of information of a cMOOC. Int. Rev. Res. Open Distance Learn. **3**, 188–217 (2014)
23. Hecking, T., Hoppe, H.U., Harrer, A.: Uncovering the structure of knowledge exchange in a MOOC discussion forum. In: Proceedings of the 2015 IEEE/ACM International Conference on Advances in Social Networks Analysis and Mining 2015, pp. 1614–1615. ACM, New York (2015)
24. Shirvani Boroujeni, M., Hecking, T., Hoppe, H.U., Dillenbourg, P.: Dynamics of MOOC Discussion Forums. In: Seventh International Learning Analytics and Knowledge Conference (LAK17) (2017)
25. Harris, J.K.: An introduction to exponential random graph modeling. Sage Publications, Thousand Oaks (2013)
26. Handcock, M., Hunter, D., Butts, C., Goodreau, S., Krivistky, P., Morris, M.: ERGM: fit, simulate and diagnose exponential-family models for networks. The Statnet Project. http://www.statnet.org (2015)

Extracting the Main Path of Historic Events from Wikipedia

Benjamin Cabrera and Barbara König

Abstract The large online encyclopedia "Wikipedia" has become a valuable information resource. However, its large size and the interconnectedness of its pages can make it easy to get lost in detail and difficult to gain a good overview of a topic. As a solution we propose a procedure to extract, summarize, and visualize large categories of historic Wikipedia articles. At the heart of this procedure we apply the method of main path analysis—originally developed for citation networks—to a modified network of linked Wikipedia articles. Beside the aggregation method itself, we describe our data mining process of the Wikipedia datasets and the considerations that guided the visualization of the article networks. Finally, we present our web app that allows to experiment with the procedure on an arbitrary Wikipedia category.

Keywords Online social networks · P2P infrastructure · Dependence · Scalability · Performance · Security

1 Introduction

Wikipedia is a large free online encyclopedia founded in the year 2001. It not only allows users the access to information but also encourages them to collaboratively work on the articles. At the time of writing the English Wikipedia contained more than five million articles and over 300,000 active users. Because of its accessibility and its large size Wikipedia has become one of the most important sources for encyclopedic knowledge in the world.

However, the richness of information in Wikipedia can also be overwhelming. It can be hard to separate relevant and irrelevant information and in addition the typical audience of Wikipedia can consist of people with a varying background knowledge.

B. Cabrera (✉) · B. König
University of Duisburg-Essen, Duisburg, Germany
e-mail: benjamin.cabrera@uni-due.de; barbara_koenig@uni-due.de

© Springer International Publishing AG, part of Springer Nature 2018
R. Alhajj et al. (eds.), *Network Intelligence Meets User Centered Social Media Networks*, Lecture Notes in Social Networks,
https://doi.org/10.1007/978-3-319-90312-5_5

While some content might be important for a detailed analysis of a topic, it might lead to an information overload for users interested only in a general overview. As a result it is necessary to provide tools that empower the users to make a choice and select relevant pieces of information.

In this article we want to focus on a particular part of the content in Wikipedia, namely the articles about history. Being interested in a historical event with some extension in time, such as the *Thirty Year's War*, the *French Revolution*, or the *Vietnam War*, one can look at the overview article of the topic or at all articles in the corresponding Wikipedia category (and their subcategories). Overview articles usually give a good introduction to a topic; however, they often consist of long texts and it may be difficult to get a grasp of the flow of singular events, such as battles and peace treaties, at one glance. Looking at all articles in a category in order to obtain an overview is often infeasible due to the sheer number of articles—thousands of articles are quite common in larger categories.

In the following we will describe a novel approach of summarizing and visualizing categories in Wikipedia. From a given Wikipedia category and its subcategories we will generate a graph whose nodes correspond to dates occurring in these articles and edges linking two dates whenever the source date precedes the target date and the corresponding articles are connected by a link (in any direction). In order to help the user navigate through the network, we propose to use the method of main path analysis to emphasize the structural backbone in a large (acyclic) graph.

In order to make it easy to try out the described approach we implemented a web app capable of visualizing the graph of a category highlighting the main path. This can be seen as a tool for *visual analytics*, which helps users to obtain deeper insights into the data using a visualizations. We encourage the readers to experiment with the web app.[1]

Related Work Mining Wikipedia data and using it for data visualization are certainly not a new topic and have been done in several, also nonscientific projects.

Part of our approach is about parsing the wiki markup files of Wikipedia to structured datasets used as input for our visualization methods. The most known related projects dealing with mining Wikipedia are DBPedia [2] and WikiData [14]. These projects aim to make Wikipedia's data machine readable and provide interfaces to the structured data. While in particular dates could have been extracted from one of these projects we opted for parsing Wikipedia dumps ourselves because the parsing infrastructure was needed for the other meta-data and we had full control of the extraction process.

Concerning the specific parsing of historic events from Wikipedia there is related work [8] offering an API to access the structured data. However, their data is limited to around 150,000 events and no extraction of relationships is performed. Beside Wikipedia as a data source, in general, TimeLine extraction is an emergent research field in NLP [11, 12]. Here, approaches focus on extracting temporal relationships

---

[1]http://www.ti.inf.uni-due.de/research/tools/wikimainpath/.

across documents, e.g., from news media using language features. In a sense [11, 12] deal with the task of building what we call the event network from a different perspective.

Beside the parsing aspect there are also existing approaches improving the search of historic events in Wikipedia. The need for tools to navigate through historic Wikipedia articles was addressed by Tiwiki [1], a tool that allows filtering articles by their dates. In our web app we offer a similar feature filter where a range of years can be specified which restricts the events under consideration. Close to our approach is also the histography project by Matan Stauer[2] which visualizes historical events from Wikipedia on a timeline. However, this approach does not provide navigational guidance by using a network approach such as main path analysis. A related project, Wikigalaxy,[3] visualizes the entire English Wikipedia as a huge graph-like structure (which resembles a galaxy) and offers support for navigating this graph. Here the focus is clearly on the visualization. Aggregation of Wikipedia data for visual analytics is also described in [5], which shows how visual analytics techniques can help Wikipedia volunteers (Wikipedians) to detect vandalism and edit wars and gather information about the trustworthiness of articles.

Finally our approach can be seen as part of the emerging field of computational history. In this field data mining and machine learning approaches are applied to large historical datasets. For example, in [4] the authors extract timelines of historical figures from Wikipedia data. However, their focus is on the mining part and less on summarization and visualization.

2   Main Path Analysis

Main path analysis describes a set of methods and algorithms on graphs which aim to reduce a potentially large graph to a much smaller path through the graph. It was first introduced by Hummon and Doreian [9] in 1989 to study the flow of scientific ideas in citation networks. Since then there have been several improvements to the method and applications to citation networks of various scientific fields [3, 13]. Recently there have also been successful applications to other types of networks [6, 7] different from the citation networks the method was originally invented for. In this section we will give a short overview of the algorithms used in main path analysis. For a more detailed explanation, see, e.g., [13].

---

[2]http://histography.io/.

[3]http://wiki.polyfra.me/.

2.1 Preliminaries

In this article we consider networks in the form of a *graph* $G = (V, E)$ consisting of a set of *vertices* V and a set of *edges* E connecting those vertices. As we will see later, main path analysis is only applicable to *directed* and *acyclic* graphs (also known as DAGs). For a directed graph we have $E \subseteq V \times V$, i.e., edges are ordered pairs of vertices. Hence there might be an edge $(u, v) \in E$ connecting $u \in V$ and $v \in V$ but not necessarily a backward edge $(v, u) \in E$. Acyclicity of a graph means that there are no cycles in the graph, i.e., no subsets $C = \{(s_1, t_1), \ldots, (s_l, t_l)\} \subseteq E$ such that $t_i = s_{i+1}$ for $1 \leq i < l$ and $t_l = s_1$. Intuitively, this means that there is no path in the graph going back to a vertex once we have left it. Additionally, as part of the algorithms described later, we will need to assign weights to edges of the graph. To this end we consider edge weights via a function $w : E \to \mathbb{N}_0$ where $w(e)$ defines the weight on an edge $e \in E$. Finally, we define a path P in G to be a sequence of vertices $P = (v_1, \ldots, v_k)$ in which no vertex appears more than once, i.e., $v_i \neq v_j$ for all $i \neq j$. The goal of our analysis will be to compute a path P representative of G—this is the *main path*.

The hidden assumption behind main path analysis is that the underlying DAG encodes some form of causal or at least sequential relationships between entities. For citation networks these relationships represent citations from newer to older scientific articles. A citation is interpreted as a reuse of ideas taken from the cited article and used in the citing article. This intuition motivates to ask for a path representing the whole graph—instead of, say, an arbitrary subgraph. A representative path, in this context, is the chain of scientific articles that lead to the current state of the research field. Additionally this intuition explains why we are considering only DAGs in the first place. On one hand, causal relationships naturally lead to directed connections because we want to identify what was the cause and what the result of an action—as opposed to only stating that entities are in some kind of relationship without stating the direction. On the other hand, the acyclic nature of the graph—at least in the case of citation networks—comes from the fact that we cannot have citations going forward in time. An author can, in theory, only cite existing articles which in turn cannot change their references anymore. The assumptions satisfied by the citation networks have to be considered for our application as well. However, before we do so in Sect. 3 we first explain some basic algorithms for computing the main path.

2.2 Main Path Computation

Main path analysis can be broken down into three steps explained in this subsection. Performed in the following order they are:

1. adding a source and a sink vertex,
2. computing edge weights guiding the upcoming path finding algorithms,

3. computing the path through the graph.

A (finite) directed, acyclic graph is guaranteed to contain vertices without incoming edges as well as vertices without outgoing edges. The main path should be as representative of the original graph as possible, so naturally we will always start the path at vertices without incoming edges and end it at ones without outgoing edges. That is because if we start at a vertex with an incoming edge we can simply add an incoming neighbor to the path and get a longer path capturing more of the original graph. The analogue holds true if the main path ends in vertices with outgoing edges. However, while there are always these potential start and end vertices, there are often multiple of them. In order to have a unique start and end vertex, which is a requirement of our algorithms in step three, we add two vertices v_s, v_e to be used as the new source (start) respectively sink (end) vertex. The new start vertex is then connected to every already existing vertex without incoming edges and every existing vertex without outgoing edges is connected to the new end vertex.

The second step of main path analysis consists of computing weights for the edges of the graph to be used as a guide by the upcoming path finding algorithms. In standard main path analysis [13] there is only one common approach to computing these weights named the *search path count (SPC)*. Taking the "flow of ideas" intuition into account the SPC weight $w(e)$ of an edge $e \in E$ is defined as the number of paths from the source vertex v_s to the sink vertex v_e that run over e. Thus edges connecting more important articles tend to have a higher SPC weight because important articles are usually cited more often and thus tend to contribute more paths to the mentioned path count. Additionally, the whole history of predecessor articles is considered, i.e., edges citing highly cited articles lead to a higher SPC weight. Since the number of paths through the graph grows potentially exponentially in the number of vertices, it is nontrivial to compute the SPC weight efficiently. For the sake of brevity we will not explain the exact method here and point to [3] for further details. To the reader familiar with network analysis the SPC weight might seem similar to stress centrality. Note, however, that stress centrality considers only shortest paths and is defined on the vertices, not the edges.

The final step of main path analysis is the actual path finding algorithm. There are different approaches for computing the main path, but they are all based on the SPC weights computed in step two. The most simple one is a greedy approach often called *local forward main path*. It starts at v_s and then adds a successor v of v_s for which the weight $w(v_s, v)$ is maximal to the path. If there are multiple choices of the same weight we simply pick one of them. The procedure is then repeated from v until we reach v_e. The resulting sequence of vertices is the main path.

For an example of how the steps of local main path analysis are performed, see Fig. 1.

A second algorithm computes the so-called *global main path*. This approach considers all paths from v_s to v_e and picks the one with the highest sum of all weights to be the main path. That is the weight of a path $p = (v_1, \ldots, v_k)$ with $v_1 = v_s$, $v_k = v_e$ is $w(p) = \sum_{i=1}^{k-1} w(v_i, v_{i+1})$ and we choose a path p for which $w(p)$ is maximal.

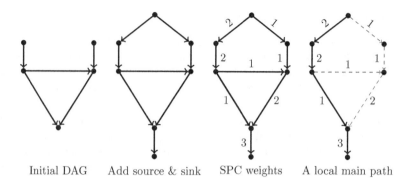

Initial DAG   Add source & sink   SPC weights   A local main path

Fig. 1 A step by step main path analysis for a basic example graph. Starting with an initial directed, acyclic graph we add a source and sink vertex. We then compute the SPC weights and run a local path finding algorithm. Note that after the second edge there are two possible choices for continuing the local path finding—we simply choose the displayed one. In contrast, the global main path would deterministically choose the other path including the third edge of weight 2

In our application area, local and global main paths have the tendency to become very long, especially since there exist categories with many thousands of articles. In some cases, the user may want to navigate such a long main path, in others, a more condensed version is required. For the latter case we use a new variant of main path analysis, the α main path algorithm. In this case, given a parameter α, the modified weight of a path $p = (v_1, \ldots, v_k)$ is $w(p) - \alpha \cdot (|p| - 1)$, i.e., the weight of the path (as defined above), minus its length (number of edges), multiplied with the factor α. This formula favors shorter paths over longer ones and the search for the path with maximal weight can still be efficiently implemented. Finding the correct α value for getting a main path of a certain length is often based on trial and error. Because the length of the main path is monotone decreasing for increasing α we can use a binary search for finding the α value that leads to a main path closest to the wanted size.

There are many further variants of local and global path finding algorithms which we do not describe in this article. They are described in detail in [13]. Note, however, that while the local main path algorithms are naturally very fast, considering all paths for a global variant can lead to exponential-time algorithms. Fortunately, there are fast algorithms also for the global methods.

3 Extracting the Main Path of Historic Events

As detailed in the introduction, it is our aim to apply the main path analysis methods described in the previous section to a use case different from citation networks. To be precise, we compute the main path of historic events based on Wikipedia datasets. To this end, we first have to be able to build a meaningful graph from the Wikipedia datasets to which a main path analysis can be applied. In this section, we first describe the general ideas about how this graph can be built and then go into some implementation details.

3.1   A Graph of Historic Events in Wikipedia

In order to build a graph, the first step is to select a group of articles that are of relevance for a particular topic. Wikipedia has a system for grouping articles on the same topic into categories. For example, there are categories on *World War II* or the *French Revolution* containing even finer subcategories such as *Battles and operations of World War II* or the *French First Republic*. A detailed overview can sometimes only be obtained by recursively searching for all subcategories and assembling all pages within subcategories. In our approach one can choose whether to perform this recursive descent (or not).

Since it is our goal to build a graph from Wikipedia articles, the first thing that comes to mind is to use the articles as nodes and hyperlinks between articles as edges. It can be argued that the links encode a meaningful relation—not only between the articles but also the article's topics. If an article about topic A references topic B, we can assume that the two topics are related. Furthermore, if an article is referenced often by articles of a certain topic, it also seems reasonable to assume that this article is more important to the topic than articles not referenced as often.

However, articles do not necessarily correspond to events. Most of the Wikipedia articles about historic events cannot be reduced to only one date. Instead, big historic events often take place in several steps so that we extract multiple dates from one article. We deal with this problem by considering each date extracted from an article as its own event labeled by the context (i.e., key, see following section) in which the date appeared in the article.

Beside that, we often extract date ranges stating that some event started at some point and carried on until a later point in time. For example looking at a topic such as the *Thirty Years' War* we have events that range over a substantial period of time while others such as the *September 11 attacks* took place on only one day. Since it is unclear how to order overlapping date ranges, we split such ranges into atomic start and end dates. Hence, the nodes of our graphs are not the articles, but the various dates. Edges are supposed to model causality and should hence point "forward in time," i.e., from an older to a newer event. More concretely, we add an edge from event a to event b whenever a precedes b and a, b are listed either in the same article or in articles which are related by a link (in either direction!). A schematic depiction can be found in Fig. 2.

We argue that an edge in the described graph encodes a causal relation between the two historic events. On the one hand, a connection in the link graph introduces some form of relation between the events that was at least important enough for the article's author to mention it. On the other hand, because the edge is pointing from the older to the more recent event, we capture that the influence of an event towards the other can only be in this direction. Together this makes it reasonable to interpret an edge from event A to B as "event A influenced—or even directly caused—event B."

The resulting graph is necessarily acyclic, making it amenable to main path analysis. A path in this graph represents a sequence of historic events from the

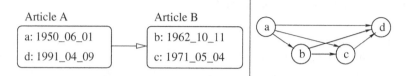

Fig. 2 Extraction of the graph of historic events (schematic depiction)

oldest to the most recent one. A main path is such a sequence summarizing the historic development in the whole graph.

Our implementation is split into two main components: the parser component and the visualization component. Both use relatively different technologies and we describe them separately in the next two sections.

3.2 Implementation of the Parser Component

As can be inferred from the previous section we need three types of structured data to execute our method:

1. the set of all Wikipedia articles with extractable dates,
2. a mapping from any category to the articles it contains (taking into account nested categories, see below),
3. the links which exist between articles.

Naturally, we want to have an automatic gathering and extraction process instead of a tedious manual extraction of articles and dates. What is difficult about this approach, however, is that Wikipedia datasets are not available as structured data. The building blocks of Wikipedia are pages, text files in a specific Wiki markup[4] that are behind each visible web page of Wikipedia. While the Wiki markup allows to format dates, positions, etc. in a specific way it does not separate them into something resembling a relational database. Thus, in order to access specific data from the pages, one has to parse the text files first. This parser component of our implementation runs on full Wikipedia XML dumps,[5] extracting the needed datasets. This enables users to analyze any category in Wikipedia containing any article with an extractable date.

The parser component consists of several C++ programs parsing the large Wikipedia XML dumps files. These programs represent the different steps necessary for extracting the required entities and serializing them to structured files where they can be used by the visualization component. The first step extracts all articles

---

[4]https://en.wikipedia.org/wiki/Wiki_markup.

[5]https://dumps.wikimedia.org/enwiki/.

with dates and all categories in Wikipedia. A category is easily recognized because its corresponding Wikipedia page title always starts with the string `Category:`, followed by the category's title. We can thus check for this string at the beginning of each page and save all the category titles to a file.

Extracting dates from articles, however, is the most challenging part of the parsing process. The reason for this is twofold. On one hand, we first have to locate a date on a page. Often dates are simply part of the text of an article or they appear sporadically on tables summarizing certain events. Fortunately, there is also a relatively reliable source of extractable dates in the form of infoboxes. Infoboxes appear on many Wikipedia articles, summarizing its content into a few key bullet points. For historic articles these often contain important dates about the event. In the Wiki Markup infoboxes are initialized by an `{{infobox` string, followed by a set of key-value pairs consisting of a label (key) and the content (value) of a bullet point. Our parsers search for infoboxes on a page and then scan the key-value pairs for keys which implicate a date in the value (e.g., `started`, `ended`, `date`, etc.).

Once we identified strings possibly containing dates we have to parse them into a common structured format. However, here lies the second problem because dates are formatted in many different ways in Wikipedia. For example, many authors will simply use a format where day is followed by month and year, while others prefer month followed by day and then year and so on. The date extraction process is made even more complicated by the fact that Wiki Markup provides special templates for dates that lead to them being displayed in a particular fashion. These also need to be parsed correctly. To incorporate all these variations we used an approach based on grammars specifying the different formats. Using the *boost.spirit*[6] library for turning the grammars into parsers we were able to reliably extract dates for most articles at a speed that allowed us to parse full Wikipedia dumps in a reasonable amount of time.[7] A schematic visualization of our date extraction scheme can be found in Fig. 3.

So far we extracted categories that can group articles into related topics and articles with detectable dates. The next parsing step now deals with getting the links needed for the edges of the graph we want to build. Links in the Wiki markup start with `[[` followed by the name of the Wiki page to which the link points and are closed by another `]]`. While in reality there are some slight variations to this format it is relatively easy to extract all links from a Wikipedia page. We call the Wikipedia page where the link occurred the source and the page it is pointing to the target. Each link is of one of the following four types:

1. source and target are articles with extractable dates,
2. source is an article with extractable date and the target is a category page,
3. source and target are category pages,
4. none of the above is true.

---

[6]http://boost-spirit.com/.

[7]Around 2 h for extracting the dates in ∼130 GB of Wikipedia pages.

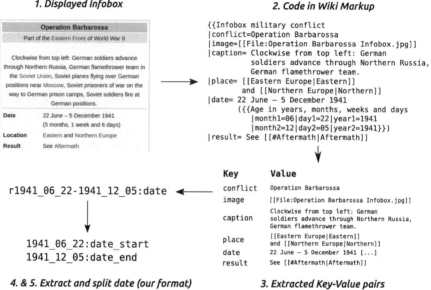

Fig. 3 The date extraction process used for extracting events with dates from a historic article. Most historic articles contain infoboxes summarizing important information including the date. Our parser detects key-value pairs likely containing dates and parses the different formats in which the date could be encoded

For the first type of links, we add an undirected link between source and target to an article network dataset. Note that we do not already split the articles into events and that we do not build a directed network yet. This will happen online for requested categories in the visualization component. For the second type of links, recall that we are selecting articles relevant for a topic by choosing a category. Links from an article to a category imply membership of the article in the category. Thus we can use the second type of links to save which articles are part of which category. The third type of links is used to solve a different problem related to categories and their containing articles. Wikipedia often uses subcategories to split large categories into smaller ones. For example, a category such as *World War II* might have subcategories such as *Battles and operations of World War II* or *World War II resistance movements*. However, if we want to analyze the full *World War II* category we have to consider all articles that are part of a subcategory of it. To this end, we use the third type of links to build a hierarchy of categories which is later used in the visualization component to recursively build the set of articles in a category from its subcategories. Finally, the remaining links of the fourth type are not relevant for our approach and can be ignored.

3.3 Implementation of the Visualization Component

The second large part of our implementation is the visualization component. It takes the structured data from the parsing step, computes the article network with its main path, and displays the result to the user. It is implemented as a web app that can be used from any web browser and consists of a HTTP server back-end written in C++ and a JavaScript client side front-end.

Using the web app, a user is first presented with a search input which allows to search for Wikipedia categories containing certain keywords. Having extracted a list of all categories in Wikipedia the implementation of this search feature is a simple full-text search scheme we will not go into detail about. After choosing a category, every upcoming step is performed on this category.

The first task performed by the server is to gather all articles that are part of the category. To this end we have to traverse the tree-like structure describing the hierarchy of categories. We start at the selected category and add all articles belonging directly to this category to a list. Then we move to all subcategories and do the same recursively. This continues until we reach all leaves. Note that because articles might belong to multiple subcategories, we use a sorted structure to only collect distinct articles. After having a list of all articles in the category we build a list of all atomic events (compare Sect. 3.1) that are part of these articles. To this end we look at all dates we extracted for the articles and turn all of them into events. At this point we can apply some filters to the event list to throw out outliers or unfitting events. Filters provided in the app include the ability to exclude events outside of specified range of dates, excluding events related to persons or excluding events containing certain keywords.

Next, we build the event network from events in the list using the procedure described in Sect. 3.1 (cf. Fig. 2). Following the construction of the network we can apply one of the main path algorithms. Finally, the back-end creates a response to the request and sends all the above-mentioned data to the client side web app. There the network is layouted according to the dates of the events (x-coordinate) and a random y-coordinate. The main path is highlighted and some statistics on the category are displayed on the side.

4 Results

The following section contains details about the dataset parsed from a current Wikipedia dump as well as representative examples from categories with their respective main path.

The dataset used in the upcoming computations was extracted from a XML dump of the whole Wikipedia from 1st May 2017.[8] In total the XML dump consisted of 40 M Wikipedia pages at a size of 130 GB. It took all programs of the parser

---

[8]https://dumps.wikimedia.org/enwiki/20170501/.

Table 1 Categories with their respective sizes and the lengths of their local main path

Category	Timespan	#Events	#Links	Local MP
Ukrainian Crisis [NR]	2011–2017	24	118	16
Events of the French Revolution by Year [NP]	1789–1804	41	189	24
Norman Conquest of England	1001–1112	42	271	26
Thirty Years War [NP]	1618–1648	81	218	10
French Revolution [NP]	1785–1849	1261	12,264	276
European theatre of World War II [NP]	1939–1945	1712	13,867	265
World War II [NP]	1939–1945	6250	60,009	329

[NR] stands for "Not Recursive" meaning that we did not consider subcategories; [NP] abbreviates "No Persons" meaning that events about persons (birth, death, ...) were filtered out. The timespan is built from the earliest and latest events in the category and can deviate from the actual historic period

component around 4 h to run on a server with 4 x Intel(R) Xeon(R) @ 3.50 GHz (16 cores in total) and 64 GB RAM. We were able to parse around 1.7 M articles that contain at least one date leading to 2.7 M atomic historic events in our final dataset. Furthermore, we extracted 1.6 M categories and their respective containing articles. Finally, we collected around 27 M links between articles used in the construction of the event network for a requested category. The computations for one category in the visualization component never exceeded three seconds although displaying very large components in the browser can result in slower response times of the app.

In order to evaluate our method we looked at several historic categories of varying sizes for which we build the event network and computed the main paths. Table 1 shows a list of the categories, their sizes, and the number of vertices in the local main path. We can observe that categories vary considerably in size, ranging from those containing only a few events to categories such as *World War II* with thousands of events. Furthermore, there are great differences in structure. This is for example visible when comparing categories *Thirty Years War* and *Ukrainian Crisis*. Although the former has more than three times the links of the latter its local main path is significantly shorter. The reason is that *Thirty Years War* is not as well connected and consists to a larger extent of parallel paths while *Ukrainian Crisis* behaves much more like one connected component.

Figure 4 shows the small but prototypical event network of the *Events of the French Revolution by Year* category.[9] One can observe that the category is well connected—not breaking up in distinct paths. As a result the local main path runs through 24 of the 41 events. The main path contains important events such as the *Storming of the Bastille* and the establishing of the *National Convention*. However, there are also important events missing. For example, the article about the *Women's March on Versailles* is not in our dataset because it contains no infobox with a date and thus no date could be extracted. Note also that the main path does not only consist of nodes with high degree—although there is a high correlation for this small category. This argues in favor of the main path method because simply showing

---

[9]Note that in the paper we use a different layout than for the web app.

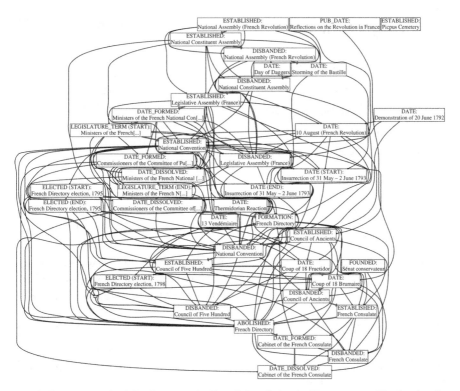

Fig. 4 Event network of the *Events of the French Revolution by Year* category. The local main path is highlighted in blue

nodes with the highest degrees would rank events differently, not considering their sequential relations. As an example, the 24 nodes with highest degree do not contain the Storming of the Bastille. This deviation between high degree and inclusion in the main path is more pronounced as the category gets larger.

Figure 5 shows the extracted local main paths for the categories *Thirty Years War* and *Ukrainian Crisis*. The one for *Thirty Years War* contains the important events *Peace of Prague*, *Battle of Nördlingen*, and the *Peace of Westphalia* that ended the war. However, the battles of Lens and Oldendorf are rather unimportant events. Also note that the article *Thirty Years War*—called like its containing category—is a summary article about the whole war. As a result it is referenced by almost every other article of the category and thus a hub of the event network. Hubs improve the connectivity of the network but, on the other hand, they usually appear on the main path which could be considered redundant because the main path itself tries to summarize a category. Another interesting observation can be made at the end of the *Thirty Years War* main path. The war is considered to end with the signing of the Peace of Westphalia. However, the summary article dates the war's end to before the end of signing of the peace treaties, with the Battle of Lens in between. The reason for this is that the treaties were signed over months and the summary

Fig. 5 Extracted local main paths from the categories *Thirty Years War* and *Ukrainian Crisis*

article's ending date is the start of the signing process while the last event of the main path marks the end of it.

We included the *Ukrainian Crisis* category to show that our method can also be applied not only to classic but also to modern history. The main path correctly starts with the Euromaidan event that started the uprising in the Ukraine, followed by the Russian intervention attempts and several treaties trying to solve the conflict.

So far the analyzed example categories were of relatively small size—although possibly still too large to be analyzed by hand. However, especially very important historic categories often contain a lot more articles. This is, in particular, the case because we aggregate all subcategories into the main category as well. As an example we picked the *European theatre of World War II* category containing over 1000 events and 13,000 links even after removing persons and events not between 1939 and 1945. The local main path contains 265 nodes which is only 15% of the original nodes but is still long in absolute terms. As a solution we can apply the mentioned α main path algorithm explained earlier where α is a parameter that can be used to influence the length of the path. Figure 6 shows α main paths for *European theatre of World War II* with varying values for α. As described in Sect. 2.2 we used a binary search to find α values for maximal path lengths $l = 10, 15, 25$. Overall we observe that some important events like the Invasion of Poland appear in all three main paths. With growing path length less important events are added. However, for instance the Normany landing appears only in the longest main path although it could be argued that this event would be important enough to already appear in the smallest one.

$l = 10$

> FIRST: Carpathian Ruthenia during World War II
> SECOND: Slovak Republic (1939–1945)
> DATE: Jabłonków Incident
> DATE (START): Invasion of Poland
> DATE: Adlertag
> DATE: The Blitz
> DATE (START): Balkan Campaign (World War II)
> DATE (START): Greco-Italian War
> DATE (END): World War II in Yugoslavia
> DATE (END): Battle of Poljana

$l = 15$

> FIRST: Carpathian Ruthenia during World War II
> FOURTH: Carpathian Ruthenia during World War II
> SECOND: Slovak Republic (1939–1945)
> DATE: Jabłonków Incident
> DATE: Gleiwitz incident
> DATE (START): Invasion of Poland
> DATE: Adlertag
> DATE: The Blitz
> DATE (START): Balkan Campaign (World War II)
> DATE (START): Greco-Italian War
> DATE (END): Battle of Crete
> DATE (END): Balkan Campaign (World War II)
> DATE (START): Operation Barbarossa
> DATE: Eastern Front (World War II)
> END_DATE: I Corps (Czechoslovakia)

$l = 24$

> FIRST: Carpathian Ruthenia during World War II
> SECOND (START): Carpathian Ruthenia during World War II
> FOURTH: Carpathian Ruthenia during World War II
> SECOND: Slovak Republic (1939–1945)
> DATE: Jabłonków Incident
> DATE: Gleiwitz incident
> DATE (START): Invasion of Poland
> DATE: Adlertag
> DATE: The Blitz
> DATE (START): Balkan Campaign (World War II)
> DATE (START): Greco-Italian War
> DATE (END): Battle of Crete
> DATE (END): Balkan Campaign (World War II)
> DATE (START): Operation Barbarossa
> DATE: Eastern Front (World War II)
> DATE (END): Operation Polyarnaya Zvezda
> DATE: Battle of Kursk
> DATE (START): Allied invasion of Sicily
> DATE: Operation Slapstick
> DATE (END): Allied invasion of Italy
> DATE (START): Bernhardt Line
> DATE (END): Battle of Monte Cassino
> DATE: Normandy landings
> DATES (END): 21st Panzer Division (Wehrmacht)

Fig. 6 Main paths extracted from the *European theatre of World War II* category using the alpha main path algorithm to enforce certain maximal path lengths *l*

5 Conclusion

We have presented a procedure to mine data from Wikipedia in order to provide a visual user aid in form of a main path of historic events. Applying the method to various categories shows that meaningful main paths are extracted containing the main events of the category.

Naturally, our approach has some limitations. For instance, it is unclear whether links between articles encode causality in all cases. It is something we assumed for our analysis and which is intuitively what a main path is describing. Another problem can, in a few instances, be the nesting of categories within Wikipedia. Some categories contain subcategories that are either completely unrelated or only remotely related to the topic. For instance, the *Vietnam War* category contains a subcategory *Counterculture of the 1960s* which contains articles such as *Abortion law* and *Psychedelic rock*.

On another note it is not always clear which main path algorithm yields the best results. The local algorithm might, due to its local decision, not always be robust with respect to small changes in the underlying graph, which favors the—still very efficient—global algorithm. Since main paths in our examples have a tendency of being quite long, the α main path algorithm helps to create short paths. In that case the question of choosing the "right" α value remains.

Future improvements could be applied to the visualization component's graph layouting algorithm. So far events are plotted on a time x-axis and randomly placed on the y-axis. As an improvement one could use a variant of a force-directed layouting algorithm [10] that places events by minimizing overlap. It would also be possible to perform an extended evaluation of our approach by letting historians judge the main paths and visualizations and their representative quality.

Acknowledgements We would like to thank Ulrich Hoppe and Stephanie Große for interesting and stimulating discussions on the topics of the paper. In addition we would like to thank Issai Zaks for his help with the servers and overall technical support. Finally, we have to thank our anonymous reviewers providing some very valuable feedback and missed references to related work. This work was supported by the Deutsche Forschungsgemeinschaft (DFG) under grant GRK 2167, Research Training Group "User-Centred Social Media."

References

1. Agarwal, P., Strötgen, J.: Tiwiki: searching wikipedia with temporal constraints. In: Proceedings of the 26th International Conference on World Wide Web Companion, Perth, April 3–7, 2017, pp. 1595–1600 (2017)
2. Auer, S., et al.: DBpedia: a nucleus for a web of open data. In: The Semantic Web, 6th International Semantic Web Conference, 2nd Asian Semantic Web Conference, ISWC 2007 + ASWC 2007, Busan, November 11–15, 2007, pp. 722–735 (2007)
3. Batagelj, V.: Efficient algorithms for citation network analysis. In: CoRR cs.DL/0309023 (2003)
4. Bauer, S., Clark, S., Graepel, T.: Learning to identify historical figures for timeline creation from wikipedia articles. In: Social Informatics - SocInfo 2014 International Workshops, Barcelona, November 11, 2014, Revised Selected Papers, pp. 234–243 (2014)
5. Boukhelifa, N., Chevalier, F., Fekete, J.-D.: Real-time aggregation of wikipedia data for visual analytics. In: Proceedings of VAST '10 (Visual Analytics Science and Technology), pp. 147–154. IEEE, New York (2010)
6. Chuang, T.C., et al.: The main paths of medical tourism: from transplantation to beautification. Tour. Manag. **45**, 49–58 (2014)
7. Halatchliyski, I., et al.: Analyzing the flow of ideas and profiles of contributors in an open learning community. In: Prof. of LAK '13 (Conference on Learning Analytics and Knowledge), pp. 66–74 (2013)
8. Hienert, D., Luciano, F.: Extraction of historical events from wikipedia. In: The Semantic Web: ESWC 2012 Satellite Events - ESWC 2012 Satellite Events, Heraklion, Crete, May 27–31, 2012. Revised Selected Papers, pp. 16–28 (2012)

9. Hummon, N.P., Doreian, P.: Connectivity in a citation network: the development of DNA theory. Soc. Netw. **11**(1), 39–63 (1989)
10. Kobourov, S.G.: Spring embedders and force directed graph drawing algorithms. In: CoRR abs/1201.3011 (2012)
11. Kolomiyets, O., Bethard, S., Moens, M.-F.: Extracting narrative timelines as temporal dependency structures. In: Proceedings of the 50th Annual Meeting of the Association for Computational Linguistics: Long Papers - Volume 1, ACL '12, pp. 88–97. Association for Computational Linguistics Jeju Island (2012)
12. Laparra, E., et al.: Multilingual and cross-lingual timeline extraction. In: CoRR abs/1702.00700 (2017)
13. Liu, J.S., Lu, L.Y.Y.: An integrated approach for main path analysis: development of the Hirsch index as an example. J. Am. Soc. Inf. Sci. Technol. **63**(3), 528–542 (2012)
14. Vrandecic, D., Krötzsch, M.: Wikidata: a free collaborative knowledgebase. Commun. ACM **57**(10), 78–85 (2014)

Identifying Accelerators of Information Diffusion Across Social Media Channels

Tobias Hecking, Laura Steinert, Simon Leßmann, Víctor H. Masías, and H. Ulrich Hoppe

Abstract This paper addresses information diffusion across different social media channels. Hereon, the time delay during the diffusion process is taken into account with a new measure. This measure can be used to identify and characterize important contributions and their interdependencies in social media according to their ability to accelerate the diffusion of information. This can also be used to find the main path of fast information diffusion. The utility of the introduced approach is demonstrated in two case studies collected with a new sampling technique.

Keywords Social media · Katz centrality · Information diffusion · Cross-media behavior · Main path analysis

1 Introduction

Online social media facilitate the spread of information among a vast amount of possible recipients on a short time scale. This is of high interest for information and knowledge management. Most existing research has analyzed information diffusion via a single media channel such as Twitter. However, information is usually spread via various channels. An example of such a media-crossing propagation is given in Fig. 1.

Here, a phrase from a concert transcript on a Bob Dylan homepage was incorporated into Bob Dylan's Wikipedia article. When CNN reported Bob Dylan's win of the literature Nobel prize, the Wikipedia article was updated. The new article revision also contains the previous revision's information and is therefore linked to it. This redundancy is shown as a dashed edge in Fig. 1. The Nobel prize news was further propagated via Twitter as a third channel.

T. Hecking (✉) · L. Steinert · S. Leßmann · V. H. Masías · H. U. Hoppe
University of Duisburg-Essen, Duisburg, Germany
e-mail: hecking@collide.info; steinert@collide.info; masias@collide.info; hoppe@collide.info

© Springer International Publishing AG, part of Springer Nature 2018   83
R. Alhajj et al. (eds.), *Network Intelligence Meets User Centered Social Media
Networks*, Lecture Notes in Social Networks,
https://doi.org/10.1007/978-3-319-90312-5_6

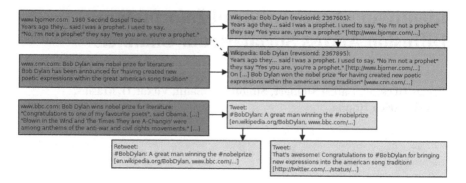

Fig. 1 Example of interdependencies between contributions from different media channels, arrows indicate retweets, revisions, and URL referencing in the direction of the information flow (cited to citing); color corresponds to the media type—Wikipedia (green), Twitter (blue), and webpages (purple)

This example motivates our research on methods to investigate the role of different contributions and media types in dissemination processes. A special focus is on temporal aspects when it comes to influence assessment. The three main contributions of this paper are as follows: (1) A novel retrieval approach to gather data from heterogeneous online media based on predefined and dynamically adapted key phrases as well as hyperlinks. (2) Definition of a new measure which allows to model influence and interdependencies between contributions by implicitly taking time into account. (3) Identification of the *main path* of information diffusion that incorporates the time-sensitive measure.

The next section discusses the background and related work. Afterwards, we describe the data collection in Sect. 3. In Sect. 4, the new time-sensitive influence measure is introduced and its application to two example case studies is presented thereafter. Section 6 discusses the findings and concludes the paper.

2 Background and Related Work

Prior to the analysis of information diffusion, the notion of an "information item" has to be clarified. Spreading information items can be modelled based on text analysis, for example, hashtags [17], n-grams, or single terms [2, 8]. Alternatively, information items can be resource-based, i.e. referring to concrete resources visible as URLs [4, 6].

Resource-based modelling is least ambiguous, however, an exclusive concentration on URL sharing can be too narrow. As an intermediate between the topic based and resource based approaches, the meme tracker algorithm [13] aims to identify multiple occurrences of different variations of short textual phrases across blog posts, which inspired the approach used in this paper.

Social influence is closely related to information diffusion. The famous hypothesis of "two-step flow of communication" [12] was among the first sociological theories that explicitly takes into account the interdependence between mass media and interpersonal communication. It says that the flow of ideas from mass-media to the population is mediated by opinion leaders who influence a large audience. This theory has further been refined and operationalized in computational models [18] with the result that the existence of "influentials" alone does not shape the information cascades observed in the real-world.

Information diffusion also received attention in the scientometrics community. Citation networks of scientific publications can be considered as prototypical cases of diffusion and influence among knowledge items. Path analyses and influence assessment are used to investigate the dissemination of scientific knowledge [7, 10]. However, techniques from the scientometrics community have not been widely adopted in the research on information diffusion in online media. Among the few exceptions is a work by Halatchliyski et al. [9] in which main path analysis [10] is applied in the context of Wikipedia.

Since the rise of social networking platforms and microblogging services, spreading cascades of pieces of information have become more explicit in social media. These cascades can be reconstructed from the underlying friendship or follower graph, direct mentioning, and visible forwarding [5, 6, 16]. However, apart from a few studies such as [1], the observable interdependencies between different information channels is widely unexplored and often relies on simulation due to the lack of empirical data (see [15] for an overview).

3   Data Collection and Preprocessing

3.1   *Network and Subgraph Extraction*

In our data collection, we focus on explicit relationships across and within multiple media types. Herein, we identify information items via the usage of similar phrases, similar to Leskovec et al. [13]. The collection starts with a search query to the Twitter streaming API[1] using a static set of search terms. Additionally, dynamic parameters are computed every 15 min from the most frequently used Twitter hashtags, usernames, and keywords in the collected data. Thus, the harvesting tool adapts to incomplete search seeds. Additionally, Wikipedia's recent changes stream[2] is frequently analyzed for article updates that meet the search criteria. Whenever a URL is found in a contribution, it is accessed via its URL and parsed for parts meeting the search criteria. If the relevant content contains another URL, this is also crawled. This recursive crawling is limited to a depth of five. All extracted

---

[1]https://dev.twitter.com/streaming/overview, as of 05/18/17.

[2]https://www.mediawiki.org/wiki/API:Recent_changes_stream, as of 05/18/17.

Table 1 Parameters used in the data harvesting and subgraph extraction

Case study	Static search terms	Subgraph parameters
Bob Dylan (Oct. 12–17, 2016)	nobel, #nobel, Nobelprize, #NobelPrize, NobelPrize, nobel prize, #bobdylan, Bob Dylan	Knockin on Heavens Door [sic!]
Schiaparelli (Oct. 19–22, 2016)	#EuropeanSpaceAgency, #ESA, European Space Agency, esa, ESA, Exomars, #Exomars, Mars, Schiaparelli, #Schiaparelli, #Mars	#ESA, esa, European Space Agency, #Schiaparelli, #ExoMars

Table 2 Extracted subgraphs before (after) data cleaning

Case study name	Description	Number of nodes				
		Twitter	Facebook	Wikipedia	Youtube	Webpages
Bob Dylan	Bob Dylan wins Literature Nobel prize (Oct. 13)	1724 (1101)	20 (5)	79 (2)	22 (9)	598 (176)
Schiaparelli	ESA's Exomars mission, failed landing of the Schiaparelli probe on Mars (Oct. 19)	4433 (4343)	9 (5)	186 (24)	22 (8)	513 (111)

contributions are stored in a database, from which a graph can be extracted in which a node represents a contribution. Two nodes are connected via a directed edge if (1) one references the other via a URL, (2) one node is the Retweet of the other node, or (3) both are consecutive revisions of the same webpage or Wikipedia article. The edge direction follows the information flow. In case of revision nodes, only references not contained in the preceding revision are included.

For each dataset one subgraph was extracted. Hereon, all textual artifacts in the dataset that contain a specific phrase, keyword or were posted by a specific user were extracted and added to the subgraph. In this context, a phrase is defined as a sequence of words enclosed in quotation marks. The phrase identification also considered slightly altered versions of phrases by using a fuzzy search [13]. It is possible that nodes containing a piece of information are linked via a path along nodes not containing this information explicitly. As the information may have been diffused along such a path, our approach tries to add paths and the nodes along them that link nodes of the subgraph. Tables 1 and 2 describe the datasets analyzed in this paper.

3.2 Data Cleaning

Before the analysis was conducted, four steps of data cleaning were performed.

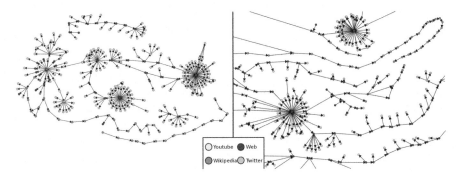

Fig. 2 Parts of the Bob Dylan (left) and Schiaparelli datasets (right) before preprocessing, containing multiple revision chains. The node colors encode the media channel

- **Missing timestamps** The timestamp of a webpage's creation or update is usually missing. As a proxy we use the time of its first reference.
- **Merging** Different URLs can refer to the same content, e.g. in a mobile and a *normal* version. We merge such duplicates. Additionally, some webpages are updated, e.g. Wikipedia articles or news feeds. This often results in long revision chains (s. Fig. 2). These falsely indicate long paths of information uptake. Therefore, we merge all revisions of a webpage. The timestamp of the merged node is set as the minimal timestamp of all merged nodes.
- **Cycle elimination** In order to apply main path analysis, the network needs to be acyclic. If all nodes of a cycle bear the same timestamp, they are merged. Otherwise, the edges that violate time constraints are deleted.
- **Bots** To filter bot posts, we use two criteria: (1) They reply to other contributions within less than 0.2 s and (2) they are not referenced. The latter criterion ensures the preservation of diffusion chains. Note that the first reference of a webpage can be mistaken for a bot message due to our timestamp proxy for webpages. However, such a contribution is only deleted if it is not referenced by others, making such a deletion relatively uncritical.

4 Time-Respecting Influence Measure

4.1 Acceleration Coefficient

An influential contribution should be (1) taken up by many influential contributions (2) in a short time. The first requirement is recursive, suggesting to use the Katz centrality [11]. It measures the influence of a node via the number of paths that exist between this node and all other nodes in the network. Thereby, longer paths are penalized by a dumping parameter α, as shown in Eq. (1).

$$\mathbf{K} = \sum_{l=1}^{\infty} \alpha^l \cdot A^l = (I - \alpha \cdot A)^{-1} - I \tag{1}$$

For the second requirement, we integrate time implicitly into the model. The timestamp difference between two connected contributions i and j is from now on referred to as *latency* $\lambda(i, j)$. Since the Katz centrality can be calculated using weighted adjacency matrices, we use the inverse of the latency of two connected contributions as a weight. Thus, the faster the takeup, the higher the weight.

Intuitively, the importance of an increase or decrease of the latency by a specific amount depends on the magnitude of the latency. For example, a latency decrease of 1 min has a larger impact if the initial latency is 2 min compared to the case where the initial latency is 2 h. Therefore, the latencies are normalized such that short uptakes of information are emphasized. The raw latency (in minutes) λ is transformed using a sigmoid function to a value in [0; 1], resulting in $\lambda_{norm}(i, j)$ given in Eq. (2).

$$\lambda_{norm}(i, j) = 2 \cdot \left(\frac{1}{1 + e^{\frac{-2 \cdot \lambda(i,j)}{\mu(i,j)}}} - 0.5 \right) \tag{2}$$

This function maps a raw latency $\lambda(i, j)$ equal to the median raw latency $\mu(i, j)$ to 0.75. The $2 \cdot (\cdots - 0.5)$ in the transformation are needed to map the function to the interval [0; 1]. Replacing the unweighted adjacency matrix A with the weighted adjacency matrix A' having elements $a'_{i,j} = a_{i,j} \cdot (1 - \lambda_{norm}(i, j))$ in the calculation of the Katz matrix K gives the temporally weighted Katz matrix K'. Consequently, the values of K', $k'_{i,j}$, are high if there exist many information pathways between i and j passing only low latency edges. Nodes that have a high temporally weighted Katz centrality with many other nodes can be considered as "accelerators" that trigger the quick takeup of information by many others. This *acceleration coefficient* (ACC) of a node i can be calculated as the sum of row i of K'.

4.2 Main Path Analysis Incorporating Time

The Katz matrix of a directed network K and its weighted version K' can be used to assess the importance of edges in the underlying diffusion process. As described in Sect. 2, the main path analysis [10] uses measures of connectivity for the analysis of the diffusion of ideas in citation networks. To this end, the influence of an older publication on a more recent one is assessed by the number of paths between them when citations are modeled as edges directed according to the flow of information, i.e. cited to citing publication. Accordingly, the importance of a citation edge is based on the number of nodes that it connects in the network. In directed acyclic graphs (DAGs) like citation networks, this corresponds to the extent an edge is stressed when information flows from source nodes (early publications) to sink nodes (latest publications).

The idea of main path analysis is taken up in the following to identify information pathways that are characterized by high traversal counts in search paths and low latency. Main path's *node pair projection count* (NPPC) weight considers all paths between all node pairs an edge occurs on [10]. In contrast, main path's *search path count* [3] only uses source-sink paths. Therefore, the NPPC weight puts more emphasis on edges that occur in the "middle" of the graph. These "middle edges" are of particular interest in the context of our analysis since they indicate indirect uptakes of source information.

We modify the NPPC weight such that edges with low latency get higher weights than edges with high latency. This altered weight called $\text{NPPC}_{\text{katz}}$ can be computed as the cross-product of row and column sums of the K' matrix, as shown in Eq. (3).

$$\text{NPPC}_{\text{katz}} = (K' \times \vec{I}) \times (\vec{I} \times K') \cdot A \qquad (3)$$

Hereon, $\vec{I}$ denotes the 1-vector. The edge weights are used to find the *global* main path [14] by applying the Bellman-Ford algorithm for finding the shortest paths between s and t. For this, all edge weights are treated as negative values.

5   Analysis of Case Studies

5.1   *Identification of Accelerators*

Figure 3 shows the density plot of ACC for the Bob Dylan case study but the ACC distribution shows the same pattern also for the other dataset. As can be seen, the value of α only stretches or compresses the distribution, but otherwise has little effect. An almost perfect Pearson correlation ($r \approx 1$, $p < 0.01$) between the ACCs with different values of α also indicates that the α parameter is only a scaling factor.

The density distribution is extremely skewed towards the left-hand side, which reveals that few nodes with extreme acceleration values exist while most have very low acceleration values. Whether the distribution of the ACC is scale-free after the first peak needs to be verified in future studies. In the *Schiaparelli* dataset, the top accelerators are mainly tweets of big news agencies or the European Space agency and information web pages. News agencies can also be found as authors among the top contributions in the *Bob Dylan* dataset, but ordinary Twitter users can be found additionally.

The ACC also correlates moderately with outdegree ($0.63 \leq r \leq 0.71$, $p < 0.01$) in both datasets. It correlates slightly negatively with indegree, but this correlation is not always statistically significant, depending on the used dataset and the α value. The acceleration coefficient can further be used to assess the importance of different media types for the dissemination of a news story across channels.

Fig. 3 The density plot of ACCs for various α values of all nodes in the Bob Dylan case study without source nodes

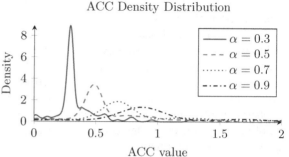

Table 3 shows the average and maximal ACC scores for the different media channels present in the two case studies. In this analysis, nodes with an indegree of 0—i.e., sources—were omitted, because such nodes do not qualify as accelerators.

It can be seen from Table 3 that the difference between the average and the maximum acceleration coefficient is mostly very high. This corresponds to the skewed distribution of the acceleration coefficient. In general, it can be said that in both datasets the strongest accelerators can be found among the web pages and tweets. One reason is that web pages usually contain much information which can easily be shared by disseminating URLs. However, the lower average ACC value for web pages indicates that this is only the case for a limited set of pages. This phenomenon is even more extreme for Twitter where only some tweets function as vehicles to distribute information among a large audience.

Table 3 Values for the ACC with $\alpha = 0.7$ of nodes with a non-zero indegree

Case study	Web	Tweet	Retweet	Youtube	Facebook	Wikipedia
Bob Dylan						
Avg.	0.29	0.10	0.00	0.99	0.23	–
Max.	4.36	2.94	0.00	1.98	0.69	–
Schiaparelli						
Avg.	7.97	0.33	0.00	1.79	0.89	0.03
Max.	449.41	136.24	0.66	3.94	1.88	0.70

5.2 Main Path Analysis of Information Diffusion

To find the path along which information was diffused most effectively in all case studies, we used the above described $NPPC_{katz}$ measure. Figure 4 shows the global main paths of the two analyzed case studies. Visualized is a two-mode network of contributions (circles) and the corresponding authors (rectangles). Note that the ACC calculation and main path analysis were performed on the 1-mode contribution graph only, i.e. excluding users.

Fig. 4 The main paths (bold edges) extracted from the contribution networks and corresponding authors; Relations between different users are indicated with dashed undirected edges; Wi: Wikipedia, We: Homepage, B: Blog, T: Twitter (* identical). (**a**) Bob Dylan case study. (**b**) Schiaparelli case study

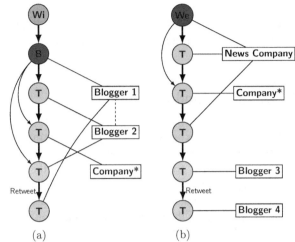

(a)   (b)

Table 4 Summary of the main paths found in the two case studies

Case study	Description
Bob Dylan	Bloggers 1, 2 are related; posts by blogger 2 appeared on blogger 1's blog
	Blogger 1's blog contains posts from arts and creativity
	Bloggers 1, 2 are active on multiple media channels, incl. Blogspot and Twitter
	The company is tweeting a lot about news, possibly as indirect advertisement
	Some users have multiple contributions on the main path
	Main path is spread via several media types
Schiaparelli	Similar patterns as in other case study
	Diffusion of information more driven by official accounts
	Dataset is much more connected than that of other case study
	3 out of 6 main path contributions come from a single major news company
	The same company found on the main path of the other case study is present
	Blogger 3 has active blog covering multiple topics
	Blogger 4's blog is inactive

The labels of the circular nodes refer to the media channel through which the information was provided. In this, *T* stands for Twitter, *B* for a blog, *Wi* for Wikipedia, and *We* for a homepage (web). Note that the type *blog* is not listed in Table 2, but is contained in *Webpages* and was manually annotated for the main paths. User labels indicate their manual classification as (news) company or blogger. Even though some users only appear on Twitter in the detected main path, blogs by users with the same name were found online. Therefore, these users were labeled as bloggers. The found main paths and the underlying users are described in Table 4.

It is interesting to note that all the users with posts on the main path were active in various media channels. In this, Wikipedia is excluded as the user responsible for an article change is not included in the datasets. Furthermore, some users have multiple posts on the main path, stressing their importance for the fast information diffusion. Overall, the media channel is switched three times in the identified main paths. Herein, the references that cross media channels link posts by the same or a closely related author in two out of three times.

6 Discussion and Conclusion

In this paper we discussed the analysis of information diffusion in online media across different platforms, an underexplored research topic. Moreover, the acceleration coefficient was presented which measures the influence of nodes in diffusion networks by incorporating a notion of time as well as reachability.

It was shown that most of the contributions in the investigated diffusion networks are not important for the spreading of information, having a very low acceleration coefficient. Moreover, the acceleration of information takeups in diffusion processes mainly relies on very few nodes. The main paths contain few users, often multiple times. These users are often active in various media channels. The references crossing different media types are often done by such users by referring to their own content on a different platform.

It was found that only a very small fraction of all contributions is followed by quick takeup sequences. This is in line with the influential hypothesis [18]. In this aspect, influence can not only be assessed locally for each node by the numbers of direct neighbors. Moreover, the influence of an individual contribution has to be seen in the context of initiation of diffusion paths over low latency edges that constitute the main path of information diffusion.

Acknowledgements This work was partially supported by the Deutsche Forschungsgemeinschaft (DFG) under grant No. GRK 2167, Research Training Group "User-Centred Social Media."

References

1. Agarwal, N., Kumar, S., Gao, H., Zafarani, R., Liu, H.: Analyzing behavior of the influentials across social media. In: Behavior Computing, pp. 3–19. Springer, Berlin (2012)
2. Aiello, L.M., Petkos, G., Martin, C., Corney, D., Papadopoulos, S., Skraba, R., Göker, A., Kompatsiaris, I., Jaimes, A.: Sensing trending topics in twitter. IEEE Trans. Multimedia **15**(6), 1268–1282 (2013)
3. Batagelj, V.: Efficient algorithms for citation network analysis. CoRR cs.DL/0309023 (2003)
4. Cao, C., Caverlee, J., Lee, K., Ge, H., Chung, J.: Organic or organized? Exploring url sharing behavior. In: CIKM'15, pp. 513–522 (2015)
5. Cogan, P., Andrews, M., Bradonjic, M., Kennedy, W.S., Sala, A., Tucci, G.: Reconstruction and analysis of twitter conversation graphs. In: HotSocial'12, pp. 25–31. ACM, New York (2012)
6. Galuba, W., Aberer, K., Chakraborty, D., Despotovic, Z., Kellerer, W.: Outtweeting the twitterers: predicting information cascades in microblogs. In: WOSN 10, pp. 3–11 (2010)
7. Gao, X., Guan, J.: Network model of knowledge diffusion. Scientometrics **90**(3), 749–762 (2012)

8. Guille, A., Hacid, H., Favre, C., Zighed, D.A.: Information diffusion in online social networks: a survey. ACM SIGMOD Rec. **42**(2), 17–28 (2013)
9. Halatchliyski, I., Hecking, T., Goehnert, T., Hoppe, H.U.: Analyzing the path of ideas and activity of contributors in an open learning community. J. Learn. Anal. **1**(2), 72–93 (2014)
10. Hummon, N.P., Doreian, P.: Connectivity in a citation network: the development of DNA theory. Soc. Netw. **11**(1), 39–63 (1989)
11. Katz, L.: A new status index derived from sociometric analysis. Psychometrika **18**(1), 39–43 (1953)
12. Katz, E., Lazarsfeld, P.F.: Personal Influence; The Part Played by People in the Flow of Mass Communications. The Free Press, New York (1955)
13. Leskovec, J., Backstrom, L., Kleinberg, J.: Meme-tracking and the dynamics of the news cycle. In: KDD'09, pp. 497–506. ACM, New York (2009)
14. Liu, J.S., Lu, L.Y.: An integrated approach for main path analysis: development of the Hirsch index as an example. J. Assoc. Inf. Sci. Technol. **63**(3), 528–542 (2012)
15. Salehi, M., Sharma, R., Marzolla, M., Magnani, M., Siyari, P., Montesi, D.: Spreading processes in multilayer networks. IEEE Trans. Netw. Sci. Eng. **2**(2), 65–83 (2015)
16. Taxidou, I., Fischer, P.M.: Online analysis of information diffusion in twitter. In: WWW'14, pp. 1313–1318. ACM, New York (2014)
17. Tsur, O., Rappoport, A.: What's in a hashtag? Content based prediction of the spread of ideas in microblogging communities. In: WSDM'12, pp. 643–652. ACM, New York (2012)
18. Watts, D.J., Dodds, P.S.: Influentials, networks, and public opinion formation. J. Consum. Res. **34**(4), 441–458 (2007)

Part III
Algorithms and Applications I

Community Aliveness: Discovering Interaction Decay Patterns in Online Social Communities

Mohammed Abufouda

Abstract Online Social Communities (OSCs) provide a medium for connecting people, sharing news, eliciting information, and finding jobs, among others. The dynamics of the interaction among the members of OSCs is not always growth dynamics. Instead, a *decay* or *inactivity* dynamics often happens, which makes an OSC obsolete. Understanding the behavior and the characteristics of the members of an inactive community helps to sustain the growth dynamics of these communities and, possibly, prevents them from being out of service. In this work, we provide two prediction models for predicting the interaction decay of community members, namely: a Simple Threshold Model (STM) and a supervised machine learning classification framework. We conducted evaluation experiments for our prediction models supported by a *ground truth* of decayed communities extracted from the StackExchange platform. The results of the experiments revealed that it is possible, with satisfactory prediction performance in terms of the F1-score and the accuracy, to predict the decay of the activity of the members of these communities using network-based attributes and network-exogenous attributes of the members. The upper bound of the prediction performance of the methods we used is 0.91 and 0.83 for the F1-score and the accuracy, respectively. These results indicate that network-based attributes are correlated with the activity of the members and that we can find decay patterns in terms of these attributes. The results also showed that the structure of the decayed communities can be used to support the alive communities by discovering inactive members.

Keywords Online social communities · Social decay · Social inactivity prediction · Inactivity patterns

M. Abufouda (✉)
Department of Computer Science, University of Kaiserslautern, Kaiserslautern, Germany
e-mail: abufouda@cs.uni-kl.de

© Springer International Publishing AG, part of Springer Nature 2018   97
R. Alhajj et al. (eds.), *Network Intelligence Meets User Centered Social Media Networks*, Lecture Notes in Social Networks,
https://doi.org/10.1007/978-3-319-90312-5_7

1 Introduction

Nowadays, Online Social Communities (OSCs) play a vital role in our daily activities. These social networks have become the main source for sharing news, connecting people, communicating, eliciting information, and finding jobs. Thus, studying the behavior and the dynamics of the members of these online platforms is crucial for sustaining these communities and maintaining the services they provide. One way to model these communities is *network* representation, where nodes represent the members of the social platform and edges represent the interactions between the members. Since the seminal works by Barabási and Réka [5] and by Watts and Strogatz [26], the field of *network science* has witnessed a huge amount of research into the dynamics of systems represented as networks. Social networks have been, and are being, studied as an example of networks that contain a lot of dynamics overtime. While a lot of social networks have been successful in sustaining their aliveness and growth dynamics, many others have experienced decay dynamics. Online social platforms such as MySpace and Friendster are now out of service due to the huge decay they have encountered, causing a massive drop of their market values.

In this work, we are interested in understanding the decay dynamics of OSCs and the interaction patterns that accompany, or possibly cause, community decay. Gaining insights into the decay interaction patterns among members of OSCs will enable us to better understand the decay process, and hence, help to provide possible actions by prolonging the life of these communities and supporting their resilience to inactivity disruptions. More precisely, we predict members who will leave a network (or become inactive) by using a Simple Threshold Model (STM) and by using a machine learning supervised binary classification framework employing network-based and exogenous features.[1] The contributions of this paper can be described as follows: (1) an exploratory data analysis of the decayed StackExchange communities and a comparison with the ones that are alive supported with a *ground truth*; (2) a Simple Threshold Model (STM) for predicting social inactivity using network-based measures or members' exogenous information; (3) a machine learning framework for predicting social leave using network-based measures and members' exogenous information; (4) guidelines for feature selection in predicting member's inactivity. The results provided insights regarding the network-based properties and also the exogenous features of inactive members that are correlated with social inactivity. These insights may help to prevent decay dynamics, to engineer resilient social networks, and to express the aliveness of OSCs.

The reminder of this paper is structured as follows. Section 2 provides a survey of related work. Section 3 gives the required definitions used in this paper. In Sect. 4, we describe the datasets used, provide the preliminary results of the exploratory data analysis we conducted, and formulate the research questions. The methods we used

---

[1]Exogenous means any information that is not based on the structure of a network.

are described in detail in Sect. 5. The results are presented in Sect. 6. In Sect. 7 the research questions are answered and the results are discussed. This paper concludes in Sect. 8, which also provides an outlook regarding the future directions.

2 Related Work

Dorogovtsev and Mendes [11] studied mathematically the decay properties of the networks and found similar characteristics of the preferential attachment provided earlier by Barabási and Réka [5]. Newman et al. [14] studied the growth dynamics of social networks between mutual friends and provided a model that showed similar characteristics as of these real networks. The growth dynamics was then studied extensively in many researches for different domains. For example, Newman [23] studied the growth dynamics, namely the clustering and preferential attachment, of scientific collaboration networks in physics and biology fields. Similarly, Bornholdt et al. [12] provided another model for simulating the growth dynamics of social network. With the availability of datasets, Barabási et al. [6] provided an empirical study on the evolution of the collaboration patterns of the scientific collaboration networks. Leskovec et al. [19] studied the growth dynamics of networks by observing some repeated patterns, namely densification laws and shrinking diameters. Backstrom et al. [4] investigated the growth dynamics of group formation and community memberships in online social networks. They provided a model for predicting when a member would join a community in a social network. A preferential attachment growth model was presented by Capocci et al. [9] to study the growth dynamics of the Wikipedia online encyclopedia. Similarly, Kossinets and Watts [17] studied the growth dynamics of a social network of students, faculty, and staff members of a university. They found that the evolution of the network was mainly affected by the network structure itself and some other external organizational structure. Kumar et al. [18] provided a large scale analysis of the social network evolution on five million members and more than ten million relationships. Their analysis revealed some structural properties of the growth process in online social networks. Ahn et al. [3] studied the growth of Myspace and Orkut, before they were permanently closed, as a real example of networks with growth dynamics. They studied the scaling behavior of the degree distribution over time for these networks and found that they have different exponents. Mislove et al. [22] studied extensively the growth of the Flicker online social network and found link formation patterns.

The aforementioned works concentrated on the *growth dynamics*. However, the dynamics of social network is not limited to growth dynamics, but also includes *decay* dynamics that may occur in the social network which leads to inactive (decayed) social network. Social network platforms like Orkut, MySpace, Friendster, and Friendfeed are now out of service after being active and growing for long time. There are few research that addressed the problem of inactivity in social networks. For example, Garcia et al. [13] studied the properties of different networks (decayed and active ones) in terms of k-core analysis. Later, Malliaros and Vazirgiannis [21] provided a method to quantify and measure the *engagement* of the members. Their measures enabled assessing the robustness of the networks over time. In a related vein, Wu et al. [27] provided a method for understanding the dynamics of social engagement of the members of the co-authorship social network of the DBLP. They showed that there was a correlation between actions of the departed members in the studied datasets. They also provided some insights regarding the properties of the members who departed the networks. Cannarella and Spechler provided an epidemic model for predicting the dynamics of the members of the Facebook [8]. The results showed that the Facebook would lose 80% of its users between 2015 and 2017, which did not happen until now. Karnstedt et al. [15], Kawale et al. [16], and Wang et al. [25] in a recent work provided prediction models for users lifespan in online social settings, also called *users churn*.

Our work is different from the previous works from two perspectives. Firstly, users churn normally has a one-to-many relationship between members and service provider. In this scenario, the social interaction, which is our main concern, is very limited. In this work, the social interaction between the users is the main concentration in the models that we provide. Secondly, our main concern is the decay of the social interaction between humans in online social networks in order to better understand the decay dynamics in online social networks.[2]

3   Preliminary Definitions

3.1   Networks and Measures

An undirected graph $G = (V, E)$ is defined as a tuple of two sets V and E, such that V is the set of nodes and the set E is the set of edges. An edge e is defined as $e = \{u, v\}$ where $u, v \in V$ and $e \in E$.

The set of neighbors of a node, $\Gamma(v)$, is the set of nodes that are connected to the node v. Table 1 shows a list of network-based measures.

3.2   Binary Classification

Given a binary variable $Y = \{True, False\}$, and a set of attributes $X = (x_1, x_2, \ldots, x_n)$ that are assumed to affect the value of Y. Then, a probabilistic supervised binary classifier in its simple forms is defined as $P(Y = True|X) = [0, 1]$ and a threshold to binarize the probability result. There are many binary classifiers and optimization techniques for finding the best model during the learning process. Describing these classifiers and learning process is beyond the goal of this paper.

---

[2]More information about our view of social decay can be found in our previous work [1, 2].

Table 1 The definitions of the used network-based measure

Measure	Description
D(v)	The *Degree* of a node v, $D(v) = |\Gamma(v)|$, is the cardinality of the set $\Gamma(v)$.
B(v)	The *Betweenness* of a node v is defined as: $B(v) = \sum_{s \in V(G)} \sum_{t \in V(G)} \frac{\sigma_{st}(v)}{\sigma_{st}}$, where $\sigma_{st}(v)$ is the number of shortest paths between the nodes s and t that includes the node v and σ_{st} is the number of all shortest paths between the nodes s and t.
$\mathcal{C}(v)$	The *Closeness* of a node v is defined as: $\mathcal{C}(v) = (\sum_{w \in V(G)} d(v, w))^{-1}$, where $d(v, w)$ is the distance between the nodes v, w.
$Core(v)$	A k-core subgraph of a graph G is the maximal subgraph such that each node has a degree at least k. The *coreness* of a node $Core(v) = k$ if the node v is in the k-core subgraph and not in the $k + 1$-core subgraph.
$E(v)$	The *Eccentricity* of a node v, $E(v)$, is the maximum distance between the node v and a node u.
Node-Cut	A *node-cut*, sometimes called articulation point, in a connected graph is a node whose removal increases the number of components in the graph.
$\mathcal{MC}(v)$	A *minimum cut* of two nodes u, v, $MinCut(u, v)$ is the minimum number of edges that are required to be removed in order to separate the two nodes. The averaged minimum cut of a node v is defined as: $\mathcal{MC}(v) = \frac{1}{n} \sum_{u \in E, u \neq v} MinCut(u, v)$, where n is the number of nodes in a graph.

Table 2 Prediction performance measures

Measure	Description
$\mathcal{P}$	The *Precision* is defined as: $\mathcal{P} = \frac{TP}{TP+FP}$.
$\mathcal{R}$	The *Recall* is defined as: $\mathcal{R} = \frac{TP}{TP+FN}$.
$\mathcal{A}$	The *Accuracy* is defined as: $\mathcal{A} = \frac{TP+TN}{TP+FN+TN+FP}$.
$F1\text{-}score$	The harmonic mean of the precision and the recall, the F1-score, is defined as $F = 2 \cdot \frac{\mathcal{P} \cdot \mathcal{R}}{\mathcal{P}+\mathcal{R}}$.

To evaluate the efficiency of a classifier, we have different measures. A true-positive (TP) happens when binary classifier classifies an instance as *True* where its real value is *True*. Similarly, we have true-negative (TN), false-positive (FP), and false-negative (FN). Based on those simple metrics, Table 2 shows the used prediction performance measures.

4 Dataset and Research Questions

4.1 The StackExchange Dataset

The StackExchange[3] is a portal that includes many question & answer websites for different focused topics. Any of these websites firstly starts as a beta community

---

[3] https://StackExchange.com/.

until it shows potential for permanent public access. However, not all of these beta communities succeed in providing the required activity and expert attraction to sustain a growth. Thus, these beta communities close. There are many examples of closed StackExchange beta communities[4] and the content generated by the users during the beta versions of them is still available. We downloaded, parsed, structurized, and analyzed a list of the closed communities in order to understand what is going during the decay (inactivity) dynamics in social networks.[5] Figure 1 shows the activity, in terms of the number of posts and the number of comments, of different websites of the StackExchange. The figure shows that the activity of the members is almost stable in the alive communities, *Statistics*, *Apple*, *Computer Science*, *German*, and *Latex* while the activity of the members is decaying in the closed communities, *Economics Literature* and the *Astronomy*. Figure 2 shows the difference between the decayed and the alive communities in terms of the active weeks. The decayed communities exhibit a significantly shorter active weeks compared to alive communities. Figure 3 shows the network representation of real data of the *Business Startups* website social network. This website was closed after a decay in its activity and we used it as a ground truth data in our experiments because it has the longest beta time which enabled us to have more meaningful snapshots overtime. The ground truth in the settings of this research is defined as the following. An observed (real) network and a predicted network are defined as $G = (V, E)$ and $G' = (V', E')$, respectively. As we are interested in predicting the nodes, we define false-negative nodes as the nodes in the set $V \setminus V'$, i.e., the set of nodes that exist in the observed network and the prediction model missed their existence. Likewise, the set $V' \setminus V$ is the set of false-positive nodes, i.e., the set of nodes that the prediction model predicted while they are not present in the observed data. Additionally, the set $V \cap V'$ is the true-positive nodes. Note that there is no true-negative as we predict only the *initial* nodes (cf. Fig. 4).

4.2   Experiment Setup and Research Questions

We define the *Members leave* problem as follows. Given a network $G_t = (V_t, E_t)$ that represents the network at time point t, and another version of the network, $G_{t'} = (V_{t'}, E_{t'})$, where $t' > t$. Then, we predict the set of nodes $V_{t'}$. During a training phase, we observe the networks G_t and G_{t_1}, then we model the properties of the nodes $V_t \setminus V_{t_1}$ in order to predict the set of nodes $V_t \setminus V_{t'}$. The set of nodes V_t are called the *initial* nodes, and any node u such that $u \in V_{t'}$ and $u \notin V_t$ is ignored for all $t < t_1 < t'$. The *inactive* members (members who left) are the set of nodes $V_t \setminus V_{t'}$. Figure 4 shows an example of the training and testing phases on exemplar networks.

---

[4]A list of the closed websites can be found here: http://bit.ly/2bVeukz.

[5]The data, the code, and the related material of this dataset are available for the public upon request.

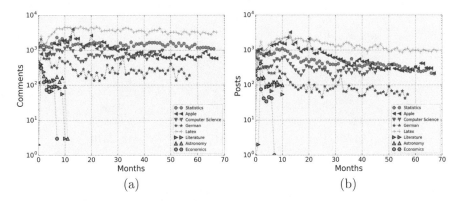

Fig. 1 The activity of members of some communities of the StackExchange websites in terms of (**a**) *comments* and (**b**) *posts* count over time. The *x*-axis represents the number of months since the launch of a website. Communities with bold markers, *Literature*, *Astronomy*, and *Economics*, were closed after the failure of their beta versions

Fig. 2 The figure shows the Cumulative distribution function (CDF) of members' active weeks. The number of active weeks is calculated as the difference between the last log-in date and the registration date of a member. The CDF is then calculated as $F(x) = P(X \le x)$. Note that the *x*-axis is log-scaled

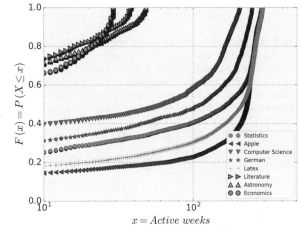

Based on that, we formulate our research questions as follows:

RQ1: *How efficient is it to predict members leaving a social community using network-based measures?* By answering this question we aim at understanding how efficient is it to use the network topology in understanding networks decay dynamics.

RQ2: *What are the network-based properties for the members who left or about to leave a community?* By answering this question we want to get insights regarding the properties of the nodes before leaving the network. Thus, networks and community maintainers can provide counter actions when a decay process starts to emerge, which enables sustaining resilient networks.

RQ3: *How helpful are the exogenous attributes in predicting members leaving?* Obtaining additional information other than the network representation is not

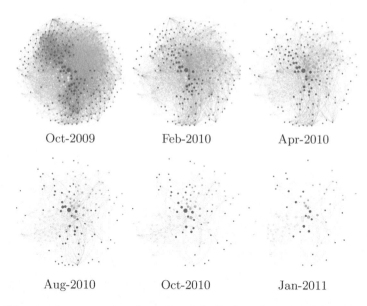

Fig. 3 The network representation of real data of the *Business Startups* website social network between Oct-2009 and Jan-2011. The networks (nodes are the members and edges represent interaction among them) were colored by clusters and having a node size directly proportional to its degree. For the sake of simplicity and for a better visualization, we restricted the networks to core members who registered-in during the first 4 months of the website. The networks in the figure show a clear decay of the number of nodes and edges

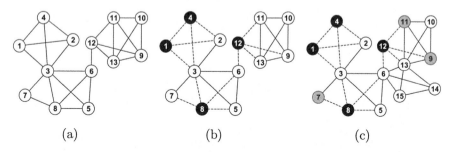

Fig. 4 A schematic illustration shows the different networks, (**a**) G_t, (**b**) G_{t_1}, and (**c**) $G_{t'}$, over time where $t < t_1 < t'$. The nodes in network G_t are *initial* nodes. The training period is performed on the time t to t_1, where the black nodes in the network G_{t_1} are the observed nodes that left the network. Then, we test on the network $G_{t'}$ where we predict the inactivity of other nodes, e.g., the grey nodes in $G_{t'}$. Note that nodes that emerge in the network $G_{t'}$ and are not in the initial node set, e.g., nodes 14 and 15, are ignored

always possible. Thus, answering this question will give us more insights on whether the network-based attributes contain sufficient information to predict the members leave. Also, we compare the prediction results performed using only the network-based attributes or only the exogenous attributes.

RQ4: *Do decayed communities embrace leave patterns that can be used to study the inactivity of communities that are alive?* The alive communities of the StackExchange may suffer from members inactivity, however, this inactivity is mitigated by the activity of new members and new discussions that support the aliveness of these communities and make them active until today. Answering this question will give us insights regarding whether there are community-independent decay patterns or not that can be used to track potential decay of the alive communities.

5 Method

In this section, we describe our method which contains the feature model that we built and used in the prediction in addition to two models for predicting *members leave*.

5.1 Features Model

We provide two types of features for the nodes of studied communities as follows:

1. *Network-based measures*: which are the values of node's attributes that are based on network measures presented in Sect. 3.1. These measures reflect how a node is connected in the network. For each node $v \in G_t$ we calculate a set ϑ_v that represents node's network-based attribute values.
2. *Exogenous attributes*: which are the values of node's non-network attributes of the members. For each $v \in G_t$ we obtain the set of attribute value θ_v. For the StackExchange dataset, these attributes include the *Upvotes* and the *Downvotes* a member received, the profile *View* counts, and the *Reputation*, among others.

In addition to the above two types, we have *Leave label*, which indicates whether a node $v \in G_t$ left the network at time t_1 or not. Thus, an instance of the feature model for a node $v \in G_t$ is defined as $I_v = \vartheta_v \cup \theta_v \cup \{True \mid False\}$. The resulted feature model is then defined as $\mathcal{M}(G_t, G_{t_1}) = \{I_v, \forall v \in G_t\}$ and used during the training phase to predict the node leave at t', where $t < t_1 < t'$.

5.2 Simple Threshold Model (STM)

We present a simple model for predicting users leave using only one attribute $i \in \vartheta_v \cup \theta_v$. The idea is that, for this attribute i we find its value for all nodes and sort them. Afterward, we find the best threshold value λ, that splits the nodes into two disjoint sets such that each element in each set has the same label, *True* or

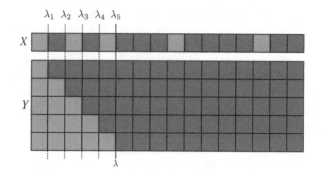

Fig. 5 The diagram shows an example on how the value of λ is computed during the training phase. The set X represents a sorted vector of the values of one attribute, say the *Betweenness* of a node, of $\mathcal{M}(G_t, G_{t_1})$ excluding the leave label for the network G_t. For the vector X, green cells mean that the node left at t_1 and the red cells mean that the node did not leave at t_1. The goal is to find a vector Y that is composed of two vectors each of them has the same value for all of its elements: True or False (in the figure green or red). The model aims at finding the best value λ of the *Betweenness* such that it maximizes the prediction performance, e.g., the F1-score. That is, the X vector is the actual labels and the Y vector is the predicted labels. The chosen $\lambda = \lambda_5$ is because the values of the F1-score are 0.3, 0.29, 0.5, 0.4, and 0.6 for λ_1, λ_2, λ_3, λ_4, and λ_5, respectively. Later values for λ, i.e., λ_q for $q > 5$, have F1-scores less than 0.6

False. The threshold value is chosen such that it maximizes one of the prediction measures provided in Sect. 3.2. Figure 5 shows a schematic diagram of this model. More formal, given sorted array of the values of an attribute i defined as $values(i)$ such that $i \in \vartheta \cup \theta$, and the corresponding leave label array. Let f be a function defined as $f : \lambda \to s$, where s is one of the prediction metrics, then the STM is defined as:

$$\arg\max_{\lambda} f(\lambda) = \{\lambda \mid \lambda \in values(i)\}. \tag{1}$$

For binary attributes, like the *articulation points*, we do not need to find the threshold λ, instead we calculate the prediction performance measure directly on the sets that contain the values of articulation points (i.e., predicted values) and the leave label (i.e., actual values).

5.3   Machine Learning Classification

With the STM, we can only benefit from the information provided by one attribute at a time. To incorporate more attributes, we used the whole feature model $\mathcal{M}$ for training and testing a supervised machine learning binary classifier to predict the *leave label*. For the evaluation, we used the same evaluation metrics presented in Sect. 3.2. We used the Support vector machines, the Logistic regression, and the Random Forests classifiers' implementation of the *scikit-learn* [24].

6 Results

6.1 Prediction Using One Attribute

In this section, we provide the results of the community decay prediction using one attribute. We also provide the results of the machine learning binary classification using one attribute only. Thus, we can compare the performance of the STM with the machine learning model. Figure 6 shows the training results of the STM. The STM performs reasonably well for most of the attributes. For example, attributes like *Betweenness*, *Coreness*, *Degree*, and *Views* show an accepted F1-score. Some other attributes like the *Upvotes* and the *Eccentricity* contain no significant information to provide good prediction as their λ was *zero*. Having trained the STM and obtained the corresponding λ for each attribute, we then predict using the attribute value at λ for different future time points. Figure 7 shows the prediction results of the STM and also the machine learning model of one attribute. To our surprise, the performance of the STM was satisfactory. For the attribute *Closeness*, the STM outperforms the machine learning model slightly with an advantage of 0.02 and

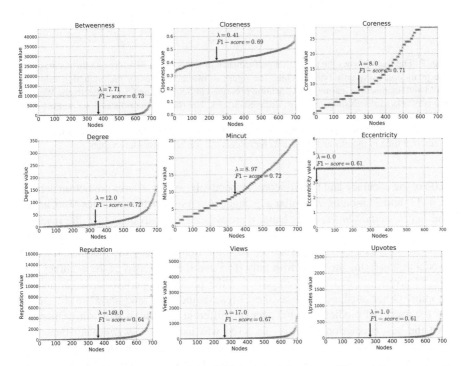

Fig. 6 The figure shows the results of the STM in the training phase. We used the networks of the *Business Startups* decayed dataset. The training period was from Nov-2009 to Jan-2010 to estimate the best λ which was used on the test period from Jan-2010 to Mar-2010. The best value of the threshold λ is shown in the figure, for each attribute associated with the value of the F1-score of the testing phase. The red and green markers indicate that the node did not leave or did, respectively. The x-axis represents the nodes ranked according to their attribute value

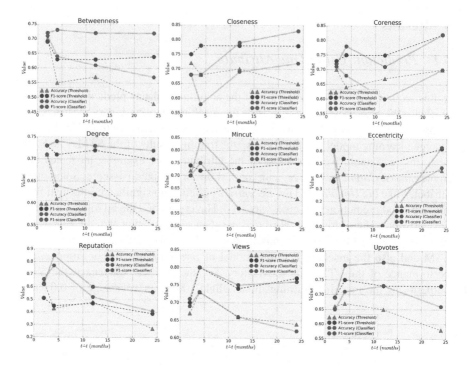

Fig. 7 The figure shows the prediction performance in terms of the F1-score and the accuracy for the prediction using only one attribute. The figure shows the results for the prediction using the STM and the machine learning classification presented in Sects. 5.2 and 5.3, respectively. We used the networks of the *Business Startups* decayed dataset. The training period was done on the period Nov-2009 to Jan-2010 to estimate the best λ which was used on the test periods 2, 4, 12, and 24 months to get more insights regarding the prediction performance. Thus, the x-axis represents the prediction time in months and the y-axis represents the prediction measure values

0.03 of the accuracy and the F1-score, respectively, averaged over prediction periods 2, 4, 12, and 24 months. A similar advantage was found on the attributes *MinCut* and the *Eccentricity*. On the other hand, there was a slight advantage for the machine learning model over the STM for the attributes *Reputation, Degree, Betweenness*, and *Upvotes*. For example, using the attribute *Betweenness*, the machine learning model outperforms the STM with only 0.05 and 0.07 in terms of the accuracy and the F1-score, respectively, averaged over the prediction periods 2, 4, 12, and 24 months. Other attributes such as the *Coreness* and the *Views* show no difference on the averaged prediction sums over the different periods. It is worth to mention that the best prediction result of the STM was using the *Coreness* attribute with prediction performance 0.83 and 0.7 for the F1-score and the accuracy, respectively, for the prediction of 24 months. Also, the machine learning model's best accuracy and the F1-score was using the attribute *Reputation* with value of 0.85 and 0.73, respectively.

6.2   Prediction Using Multiple Attributes

In this section, we provide the prediction results of our machine learning framework. We used machine learning because we may lose a lot of information when limiting our prediction to only one attribute. We emphasize here that all of the experiments performed in this section were performed on two different datasets: one for the training phase, and the other for the testing phase, which supports validity of our results and conclusions which eliminates overfitting.

Features Properties

We started by investigating the properties of the used features. Ranking features according to their importance is vital in selecting the best attributes during training the testing phases. Figure 8 shows the importance of the features used in this work. The figure shows that the *Coreness* and the *Closeness* are the most important network-based features and the *Views* and the *Reputation* are the most important exogenous information features. The information provided in this figure is valuable in selecting the best set of features. Thus, we provide different training and testing variations of the feature model $\mathcal{M}$ as follows:

1. $\mathcal{M}(\text{all})$, which uses all features.
2. $\mathcal{M}(\text{Best}_4)$, which uses the best 4 features based on Fig. 8.
3. $\mathcal{M}(\text{Best}_1)$, which uses the best feature from the network-based features and the best feature from the exogenous attributes.
4. $\mathcal{M}(\text{Best}_2)$, which uses the best two features from the network-based features and the best two features from the exogenous attributes.

Using the set of all features is not always the best choice due to some inherited properties of the machine learning classifiers. For example, some classifiers are sensitive to correlated attributes and many classifiers perform poorly with low variance attributes. Thus, we provide additional analysis of the attributes in order to better understand the features. Figure 9 shows the Pearson's correlation coefficient matrix of the attributes. Values close to -1 indicate a negative correlation, while values near to 1 indicate a positive correlation. It is preferable to feed machine learning classifiers with as more uncorrelated features as possible. We notice from Fig. 9 that the exogenous features are more correlated to each others. Also, the network-based attributes are more correlated to each other. To see that, Fig. 10 provides more information about distribution of the attributes along with one-to-one scatter plot. For example, we notice that *Coreness* vs *Degree* and *Closeness* vs *MinCut* are highly correlated.

Figure 10 shows also the distribution of each attribute in the diagonal. The features *Closeness*, *Coreness*, *Degree*, and *MinCut* have more variance than the others, which gives an interpretation on why these attributes got more importance in Fig. 8. The data shown in Fig. 6 are clearly non-linearly separable, i.e, there exist

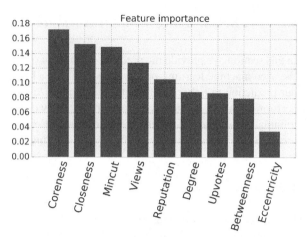

Fig. 8 The figure shows the feature importance such that $\sum_{i}^{i \in \vartheta \cup \theta} w(i) = 1$, where $w(i)$ is the feature importance. The used method for generating the importance is random forests where the importance of a feature increases whenever the split in the tree using that feature minimizes the prediction error [20]

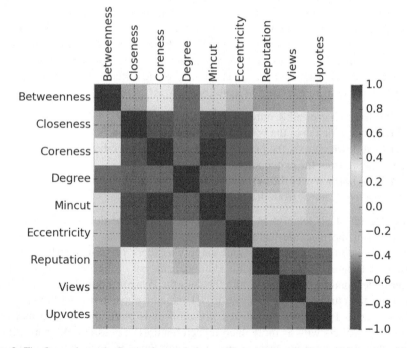

Fig. 9 The figure shows the Pearson's correlation coefficient values for the used features and it is defined as: $\rho(f_1, f_2) = \frac{Covariance(f_1, f_2)}{\sqrt{Variance(f_1) \cdot Variance(f_2)}}$, where $f_1, f_2 \in \vartheta \cup \theta$ and $\rho(f_1, f_2) \in [-1, 1]$. The data used to generate this figure is the *Business Startups* for the period between *Jan*-2010 and *Mar*-2010

no possible threshold that separates the green and the red points. The separation of the data becomes even harder when incorporating more features, like the data points in Fig. 10. In Fig. 11, we show an exemplary prediction on a 2-*d* data of the used attributes. The figure illustrates how classifiers, such as the support vector

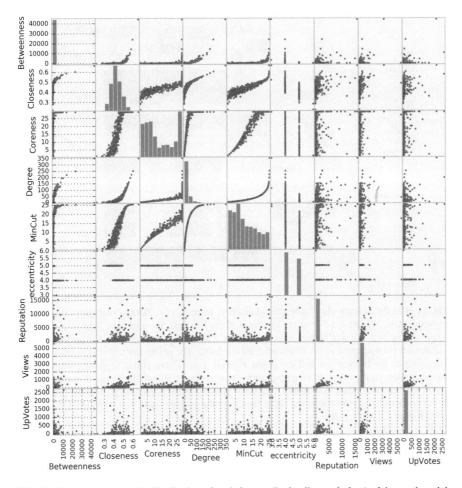

Fig. 10 The figure shows the distribution of each feature (in the diagonal plots) of the used model and also shows the correlation plot between each two attributes. The green points are nodes who left the network and the green ones are who did not. The data used to generate this figure is the *Business Startups* for the period between *Jan*-2010 and *Mar*-2010

machines, are able to provide a smooth probabilistic areas for separating points. For example, the blue and the red points of the *MinCut* vs *Closeness* are very interweaving, and the SVM managed to find good separation areas compared to the Logistic regression, Decision Trees, and Random forest. That's why we resort basically to the SVMs in the following results.[6]

---

[6]The technical details of the SVMs can be found here [10].

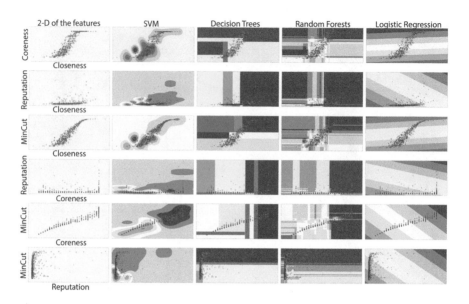

Fig. 11 The figure shows how able are the machine learning classifiers to segregate non-linearly separable data. We show a 2-*d* of the attributes of the *Business Startups*. The figures in most left column show two features' values as a scatter plot. The blue points represent the nodes who left and the red points represent the nodes who did not leave. The points marked with solid circle are the points of the training set and the points marked with '+' are points of testing set. Each of the other plots in the other columns represents how different probabilistic classifier managed to classify the corresponding 2-*d* dataset. The color gradient in the prediction plots is the probability of the prediction, i.e., the darker the color, the closer the probability to 1 or 0 (where 1 means left and 0 means did not leave)

Prediction Results

In this section, we provide the prediction results of the machine learning prediction framework. Table 3 shows the prediction results of the *Business Startups* decayed dataset using the *four* variations of the framework and for *four* time periods. The prediction results overall are satisfactory and the variation $Best_2$ gave the best prediction results over the other variations for all prediction periods. One thing to notice is that, when the prediction time grows in the prediction period, the prediction performance increases. One interpretation is that the machine learning classifiers were able to learn the leave patterns much more than the stay pattern, given that the decayed communities networks end with a disconnected network, i.e., all of its nodes had already left the network. Similar results were found for the *Literature* dataset in Table 4. The prediction results of the *Literature* dataset are higher than the dataset of *Business Startups*. After investigating the two datasets, we found the *Business Startups* community went through two phases of decay. After the first decay of the community website, there was another relaunch which was also not successful and ended with second decay. This interprets its long time span

Table 3 The table shows the prediction results of the machine learning classifier for the networks constructed from the decayed *Business Startups* community dataset

Attributes	2 Months		4 Months		12 Months		24 Months	
	$\mathcal{A}$	F1	$\mathcal{A}$	F1	$\mathcal{A}$	F1	$\mathcal{A}$	F1
$\mathcal{M}(\text{All})$	0.72	0.74	0.68	0.79	0.69	0.79	0.76	0.85
$\mathcal{M}(\text{Best}_4)$	0.73	0.74	0.81	0.72	0.68	0.77	0.72	0.82
$\mathcal{M}(\text{Best}_1)$	0.74	0.73	0.66	0.73	0.66	0.76	0.7	0.81
$\mathcal{M}(\text{Best}_2)$	0.72	0.75	0.80	0.87	0.81	0.88	0.78	0.87

The training period was from Nov-2009 to Jan-2010. The table shows the prediction for different testing sets, namely after 2, 4, 12, and 24 months. The prediction was done using different variations of the attributes model $\mathcal{M}$ as presented in Sect. 5.1

Table 4 The table shows the prediction results of the machine learning classifier for the networks constructed from the decayed *Literature* community dataset

Attributes	2 Months		4 Months		8 Months		12 Months	
	$\mathcal{A}$	F1	$\mathcal{A}$	F1	$\mathcal{A}$	F1	$\mathcal{A}$	F1
$\mathcal{M}(\text{All})$	0.68	0.77	0.83	0.91	0.82	0.9	0.88	0.94
$\mathcal{M}(\text{Best}_4)$	0.7	0.76	0.83	0.91	0.82	0.9	0.88	0.94
$\mathcal{M}(\text{Best}_1)$	0.82	0.85	0.83	0.91	0.82	0.9	0.88	0.94
$\mathcal{M}(\text{Best}_2)$	0.68	0.72	0.83	0.91	0.82	0.9	0.88	0.94

The training period was performed from Aug-2011 to Sep-2011. The table shows the prediction for different testing sets, namely after 2, 4, 12, and 24 months for different variations of the attributes model $\mathcal{M}$ presented in Sect. 5.1

compared to other decayed communities. Thus, there was a fluctuation in the activity of that community which made it harder for the classifiers to identify the real leave patterns in this community. Table 4 shows no significant difference between the *four* variations of the model, except the variation Best$_1$ which shows a slight advantage over the other variations in the prediction period 2 months.

It was tempting to test the alive communities also. Hence, we used the *Latex* alive community to predict the leave of their members also. The prediction performance for it was also satisfactory. Again, the variation Best$_1$ shows a slight advantage over the other variations, except for the period prediction over 36 months, which is considered a long time window to predict as shown in Table 5. Then, we predicted the future of the activity of the members of the active communities such as the *Latex* and the *Statistics* using the leave information of the decayed communities such as *Business Startups* and *Literature*. Table 6 shows the prediction results when training the classifiers on the datasets of decayed communities and test the classifier on the datasets of alive communities. The results suggest a better prediction performance when compared with the prediction on the same communities, such as the results in Table 5. For instance, the F1-score at time period 4 months was 0.89 when trained on a decayed dataset compared to 0.80 when trained on the *Latex* dataset itself. The prediction also shows satisfactory results for predicting the leave of the *Statistics* community when learning from the decayed communities.

Table 5 The table shows the prediction results of the machine learning classifier for the networks constructed from the **alive** "Latex" community dataset

Attributes	2 Months		4 Months		12 Months		24 Months		36 Months	
	$\mathcal{A}$	F1	$\mathcal{A}$	F1	$\mathcal{A}$	F1	$\mathcal{A}$	F1	$\mathcal{A}$	F1
$\mathcal{M}$(All)	0.71	0.7	0.7	0.71	0.73	0.79	0.65	0.74	0.6	0.73
$\mathcal{M}$(Best$_4$)	0.73	0.73	0.74	0.78	0.75	0.82	0.83	0.9	0.83	0.9
$\mathcal{M}$(Best$_1$)	0.73	0.71	0.74	0.77	0.75	0.82	0.82	0.89	0.81	0.89
$\mathcal{M}$(Best$_2$)	0.71	0.67	0.72	0.80	0.74	0.82	0.81	0.88	0.88	0.94

The train period was performed for the period Jun-2010 to Sep-2010. The table shows the prediction for different testing sets, namely after 2, 4, 12, 24, and 36 months for different variations of the attributes model $\mathcal{M}$ presented in Sect. 5.1. Being alive, the Latex community made it possible to predict using 36 months

7 Discussion

7.1 Answering the Research Questions

Discussion on RQ1 *How efficient is it to predict members leaving a social community using network-based measures?*: Based on the previous presentation of the models and the results presented in Section 6, it is clear that using network-based attributes provides good prediction performance in terms of F1-score and the accuracy. The simple prediction model showed acceptable prediction results when using only one network-based measure. The results were even better and more robust when using multiple network-based attributes for the machine learning model. However, not all of the attributes were of equal quality for decay prediction. For example, the *Eccentricity* measure was rather useless as it showed bad prediction performance using the STM. Even worse, that measure is misleading as it showed very high prediction for the 24 months using the STM, which was only the case because its initial λ was calculated as *zero*. It may be impossible to have both the network structure and exogenous information available, as one or other might be hard to collect. Thus, an interesting aspect regarding this research question is to test the prediction using only the attributes of either the network-based features or the exogenous features. Our preliminary results for this sub-question showed very close prediction performance for perdition using only one of the two types of the features. For the *Business Startups*, the accuracy and the F1-score were 0.75 and 0.83, respectively, for the network-based measures compared to 0.77 and 0.87 for the exogenous features. These results are comparable and suggest that both types of features can be used as a predictor for the decay of the communities.

Discussion on RQ2 *What are the network-based properties for the members who left or about to leave community?*: Based on Fig. 7, members with less *Betweenness*, less *MinCut*, less *Degree*, less *Closeness*, or less *Coreness* are more susceptible to becoming inactive. This conclusion is also supported by Fig. 10 and by the prediction using machine learning with one attribute and using the STM as shown

Table 6 Results of cross-datasets prediction where the training was performed on decayed communities and the test was performed on alive communities

Datasets		4 Months		8 Months	
Train on (decayed)	Test on (alive)	$\mathcal{A}$	F1	$\mathcal{A}$	F1
Business Startups	Latex	0.83	0.89	0.88	0.94
	Statistics	0.79	0.84	0.79	0.81
Literature	Latex	0.72	0.74	0.80	0.89
	Statistics	0.77	0.80	0.72	0.78

We trained the machine learning classifier on the $\mathcal{M}(Best_2)$, which provided the best results

in Fig. 7. The STM can be utilized as a decay indicator when those attributes reach the corresponding λ of the members of a community.

Discussion on *RQ3* *How helpful are the exogenous members attributes in predicting members leaving?*: The attributes used, which are based on exogenous information, also showed a potential for providing good prediction results. However, not all of these attributes were helpful. Figure 8 suggests that the network-based measures were more important than the exogenous attributes.

Discussion on *RQ4* *Do decayed communities embrace leave patterns that can be used to study the inactivity of communities that are alive?*: Interestingly, the cross-community prediction results shown in Table 6 suggest that the leave patterns are independent of the community, as we were able to predict the inactivity of a community from the information another one. Apparently, the leave patterns are universal across communities when abstracting the interaction as a network.

7.2 Threats to Validity

Network Quality The networks used in the experiments were constructed from interactions between the members of the StackExchange website, where the nodes of these networks are the members and the edges are the interactions (such as comments) among the members. To guarantee good quality of the network, we took the following steps: For each of the communities we used, a link was considered if it appeared at least one time during the training period. Other values for link persistence overtime yielded sparse network. The training period was selected depending on the number of months a community survived; for example, for the Business Startups community, the training period for constructing the network G_t was $\delta = 45$ days. We tried different values for δ. For values $\delta = 45 \pm 5$ days, there was no significant difference in the results. For larger values, e.g., $\delta = 90$ days, we got few networks that are very dense and can hardly capture any meaningful interaction patterns; for smaller values, e.g., $\delta = 5$ days, we got too many very-sparse networks. The same argument is applied for the other communities. Thus, we do not expect these design decisions to affect the internal validity of the results.

Training Quality The networks we used are decaying networks, which means that the nearer the network to the time at which community closed, the more inactive member it has. This makes prediction easier for the most recent networks. However, prediction at early time points showed satisfactory results, too, with the ratio between active and inactive members being 55:45. Although we used two different datasets (networks at two different points of time) for training and testing, we used k-fold cross-validation, with $k = 3$, to further eliminate any random chances of classification bias during the training.

8   Conclusion and Future Directions

Network-based attributes are a good representative of activity behavior in online communities. The STM, which uses only one attribute, is able to effectively predict users' inactivity. The presented method for predicting the decay of the members of online social communities gives information about the attributes of members who became inactive. Those attributes, the network-based attributes, and some other community dependent attributes can be used as indicators for the aliveness of an online community. In addition, these attributes can be used to take counter actions when inactivity behavior is detected. Such actions may include, in the context of StackExchange communities, new questions and good answer recommendations as well as additional rewards (like badges and points) for the members. One aspect of the methods contributed in this work is the computational complexity. Some network-based attributes are computationally expensive to compute, especially for large and sparse networks. However, we found that the best results were obtained from attributes that are easy to compute, like *Degree* and *Coreness* [7]. We recommend starting with the STM before using the machine learning classifiers, as the machine learning classifiers are computationally expensive for large datasets. The STM provides good indications regarding which attributes to use. The optimization of the STM is computationally easy for a sorted list; it is $\mathcal{O}(n)$ where n is the number of nodes in the graph. For 2-month prediction, the upper bounds are 0.85 and 0.82 for the F1-score and the accuracy, respectively. For the 4 months, the upper bounds are 0.91 and 0.83 for the F1-score and the accuracy, respectively. The prediction results obtained from nearer time periods to the close time cannot be generalized as the life times of the decayed communities are not equal. Future work will include the following: (1) testing network-based features other than the features presented in this work; (2) testing the method on additional datasets with decay ground truth.

Acknowledgements This research was performed as part of Mohammed Abufouda's Ph.D research supervised by Prof. Katharina A. Zweig.

References

1. Abufouda, M., Zweig, K.A.: A theoretical model for understanding the dynamics of online social networks decay. Preprint (2016). arXiv:1610.01538
2. Abufouda, M., Zweig, K.A.: Stochastic modeling of the decay dynamics of online social networks. In: Proceedings of the 8th Conference on Complex Networks, pp. 119–131 (2017)
3. Ahn, Y.-Y., Han, S., Kwak, H., Moon, S., Jeong, H.: Analysis of topological characteristics of huge online social networking services. In: Proceedings of the 16th International Conference on World Wide Web, pp. 835–844. ACM, New York (2007)
4. Backstrom, L., Huttenlocher, D., Kleinberg, J., Lan, X.: Group formation in large social networks: membership, growth, and evolution. In: Proceedings of the 12th ACM SIGKDD International Conference on Knowledge Discovery and Data Mining, pp. 44–54 (2006)
5. Barabási, A.-L., Albert, R.: Emergence of scaling in random networks. Am. Assoc. Adv. Sci. **286**(5439), 509–512 (1999)
6. Barabási, A.-L., Jeong, H., Néda, Z., Ravasz, E., Schubert, A., Vicsek, T.: Evolution of the social network of scientific collaborations. Phys. A Stati. Mech. Appl. **311**(3), 590–614 (2002)
7. Batagelj, V., Zaversnik, M.: An o (m) algorithm for cores decomposition of networks. Preprint (2003). arXiv:cs/0310049
8. Cannarella, J., Spechler, J.A.: Epidemiological modeling of online social network dynamics. Preprint (2014). arXiv:1401.4208
9. Capocci, A., et al.: Preferential attachment in the growth of social networks: the internet encyclopedia wikipedia. Phys. Rev. E **74**(3), 036116 (2006)
10. Cortes, C., Vapnik, V.: Support-vector networks. Mach. Learn. **20**(3), 273–297 (1995)
11. Dorogovtsev, S.N., Mendes, J.F.F.: Scaling behaviour of developing and decaying networks. EPL (Europhys. Lett.) **52**(1), 33 (2000)
12. Ebel, H., Davidsen, J., Bornholdt, S.: Dynamics of social networks. Complexity **8**(2), 24–27 (2002)
13. Garcia, D., Mavrodiev, P., Schweitzer, F.: Social resilience in online communities: the autopsy of friendster. In: Proceedings of the First ACM Conference on Online Social Networks, pp. 39–50. ACM, New York (2013)
14. Jin, E.M., Girvan, M., Newman, M.E.: Structure of growing social networks. Phys. Rev. E **64**(4), 046132 (2001)
15. Karnstedt, M., Rowe, M., Chan, J., Alani, H., Hayes, C.: The effect of user features on churn in social networks. In: Proceedings of the 3rd International Web Science Conference, p. 23. ACM, New York (2011)
16. Kawale, J., Pal, A., Srivastava, J.: Churn prediction in mmorpgs: a social influence based approach. In: 2009 International Conference on Computational Science and Engineering, vol. 4, pp. 423–428 (2009)
17. Kossinets, G., Watts, D.J.: Empirical analysis of an evolving social network. Science **311**(5757), 88–90 (2006)
18. Kumar, R., Novak, J., Tomkins, A.: Structure and evolution of online social networks. In: Proceedings of the 12th ACM SIGKDD International Conference on Knowledge Discovery and Data Mining, KDD, pp. 611–617. ACM, New York (2006)
19. Leskovec, J., Kleinberg, J., Faloutsos, C.: Graphs over time: densification laws, shrinking diameters and possible explanations. In: Proceedings of the 11th ACM SIGKDD International Conference on Knowledge Discovery in Data Mining, pp. 177–187. ACM, New York (2005)
20. Louppe, G., Wehenkel, L., Sutera, A., Geurts, P.: Understanding variable importances in forests of randomized trees. In: Advances in Neural Information Processing Systems, pp. 431–439 (2013)
21. Malliaros, F.D., Vazirgiannis, M.: To stay or not to stay: modeling engagement dynamics in social graphs. In: Proceedings of the 22nd ACM International Conference on Conference on Information and Knowledge Management, pp. 469–478. ACM, New York (2013)

22. Mislove, A., Koppula, H.S., Gummadi, K.P., Druschel, P., Bhattacharjee, B.: Growth of the flickr social network. In: Proceedings of the First Workshop on Online Social Networks, pp. 25–30. ACM, New York (2008)
23. Newman, M.E.: Clustering and preferential attachment in growing networks. Phys. Rev. E **64**(2), 025102 (2001)
24. Pedregosa, F., et al.: Scikit-learn: machine learning in Python. J. Mach. Learn. Res. **12**, 2825–2830 (2011)
25. Wang, Y., Guo, Y., Chen, Y.: Accurate and early prediction of user lifespan in an online video-on-demand system. In: IEEE 13th International Conference on Signal Processing (ICSP), pp. 969–974 (2016)
26. Watts, D.J., Strogatz, S.H.: Collective dynamics of small-world networks. Nature **393**(6684), 440–442 (1998)
27. Wu, S., Das Sarma, A., Fabrikant, A., Lattanzi, S., Tomkins, A.: Arrival and departure dynamics in social networks. In: Proceedings of the 6th ACM International Conference on Web Search and Data Mining, pp. 233–242. ACM, New York (2013)

Extended Feature-Driven Graph Model for Social Media Networks

Ziyaad Qasem, Tobias Hecking, Benjamin Cabrera, Marc Jansen, and H. Ulrich Hoppe

Abstract One of the major problems in social media research on networks is the lack of empirical datasets. To avoid this problem, different graph generation models have been proposed to represent real social graphs to an acceptable extent. This enables researchers to try and evaluate new methods on a large number of social media networks. The work described here aims to introduce an extended feature-driven model that provides synthetic graphs that are sufficiently representative of real retweet-based graphs generated from Twitter datasets. We have used three real retweet-based graphs introduced from three Twitter datasets, and structure-based graph metrics (statistical similarity metrics) to evaluate the performance of our extended model. Our experimental results demonstrate that our extended model stands out as a useful model to provide a highly accurate representation of the real-world graphs over other proposed feature-driven models.

Keywords Graph models · Feature-driven models · Real-world graphs · Twitter

1 Introduction

When novel methods for network analysis are developed, a major objective is to prove their validity and to study their behavior in and for different types of networks in a "what-if" analysis. However, often there is only limited availability of empirical datasets that are representative for various network configurations. For example, one might be interested in the relationship between the distribution of certain centrality measures the density in networks that resemble elementary structural

Z. Qasem (✉) · M. Jansen
Computer Science Institute, University of Applied Science Ruhr West, Bottrop, Germany
e-mail: ziyaad.qasem@hs-ruhrwest.de

T. Hecking · B. Cabrera · H. U. Hoppe
Department of Computer Science and Applied Cognitive Science, University of Duisburg-Essen, Duisburg, Germany

© Springer International Publishing AG, part of Springer Nature 2018
R. Alhajj et al. (eds.), *Network Intelligence Meets User Centered Social Media Networks*, Lecture Notes in Social Networks,
https://doi.org/10.1007/978-3-319-90312-5_8

features of Twitter networks. Gathering adequate empirical data for this purpose is extremely time consuming and sometimes even impossible such as historical or topical datasets. Thus, a typical approach is to develop models for generating synthetic networks with similar properties than a certain type of empirical networks. The goal is not to produce an exact copy of a target network (empirical network) but to create a network with similar structural features. Apart from maintaining general properties of the target network, these feature-driven models should be configurable in the sense that the structure of the output network can be adjusted based on input parameters. This enables the developers of new network measures and algorithms to analyze how new methods react on structural changes of an input network in a controlled setting.

Furthermore, generative models can also be used for hypothesis testing [25]. In this, one strategy for testing if certain phenomena, for example preferential attachment, lead to networks exhibiting specific structural properties is to build a parameterizable model and to investigate the relationship between the input parameters and the produced network structure. It is, thereby, of particular interest to find models that are easily interpretable extracting the relevant input parameters that shape the structure of a network.

This paper introduces a novel network generator for evolving directed networks that is especially be suited to simulate growing networks in domains where new nodes attach to existing ones, such as retweet networks in Twitter or post-reply networks in discussion boards. The evaluations show that elementary structural properties of empirical Twitter networks can be reproduced with considerable accuracy.

The rest of the paper is organized as follows: Sect. 2 gives an overview on relevant literature on synthetic graph generation highlighting the differences between structure-driven and feature-driven models. After that, two feature-driven models (Barabasi-Albert [1] and the Forest Fire model [9]) are described in detail in Sect. 3 since they are used for comparison with the new approach which will be described in Sect. 4. The evaluation results on fitting the proposed model to recreate structural properties of Twitter retweet networks are presented in Sects. 5 and 6. Finally, conclusions are drawn and an outlook for further research is described in Sect. 7.

2 Literature Review

In recent years, there has been considerable research in the structure of real-world networks, and in modeling these networks as the results of several random procedures. Most of the graph models try to fit graph properties such as node degree distribution, clustering coefficient, average path length, diameter, etc.

In [13, 15, 30], different random graph models have been surveyed such as: classic random graph models [4, 6, 23], small-world models [29], and preferential attachment models [1]. Sala et al. [21] have classified those graph models into three categories. Firstly, *feature-driven models* which focus on generating synthetic graphs mimicking the statistical properties of real-world graphs such as power-law degree distributions and dynamic changes in network density over long periods. Models included in this category are the Forest Fire [9] and the Barabasi-Albert [1] models. Secondly, *intent-driven models* try to imitate the real-world processes of network generation. Examples are Random Walk and Nearest Neighbor models [27]. As the third category, *structure-driven models* try to generate synthetic graphs with the same structural restrictions as real-world graphs, e.g. Kronecker [8] and dK-graph [10] models.

In social networks research, due to the lack of empirical datasets for several reasons, several research led to graph models generation to produce synthetic graphs that can be replaced by real-world graphs to conduct the social networks research on them. Barrett et al. [2] introduced a model to generate synthetic contact-based graphs imitating urban and rural regions in the US (structure-driven models). They have used several real-world datasets and social theories to produce the links in the synthetic graphs. Sala et al. [21] tried to calibrate different random graph models to imitate real friendship graphs in Facebook. They have studied six different graph models to figure out which a proper model can be used to represent the real friendship-based graphs that have been introduced from Facebook.

Our work is related to the research presented in [21] in the sense that we aim to calibrate the different graph models to generate synthetic graphs that can represent real retweet-based graphs. Through graph models calibration, we introduce a new extended graph model that has highly accurate representation of the real retweet-based graphs compared to other proposed graph models. As well as, we have used the statistical similarity metrics (as in [21]) to evaluate the accuracy of each model in our study. In this paper, we have calibrated just the feature-driven models because of the following reasons:

- The structure-driven models acquire very high costs in memory or computation. Unlike other graph models that require creation parameters, this graph models have no such parameters. They require a set of design constraints (required degree distribution, clustering coefficient, etc.), then they try through some algorithms to generate synthetic graphs that match those constraints. Thus, those models are useful, but graph generation process using such models becomes costly for large graphs as in our situation.
- The most of intent-driven models are restricted to specific activities or properties. For example, the Random Walk model simulates the behavior of friends uncovering in social media networks. In this paper, our goal is to simulate the retweet activities in social media networks.

Thus, we targeted the feature-driven models to compare them using the statistical similarity metrics with our extended model, and show that our model is useful in real retweet-based graphs representation.

3 Feature-Driven Models

Feature-driven models focus on the statistical law of the network evolution (degree distribution follows power-law), dynamic changes in network density over long periods, etc. [21]. In this section, we succinctly represent two well-known feature-driven models—the Forest Fire and the Barabasi-Albert models.

3.1 Forest Fire

Real-world social networks often show a particular dynamic evolution with characteristic effects on the density and diameter over time. Leskovec et al. [9] introduced a dynamic network model which simulates this process as a forest fire that grows by burning trees close by. In this model, the social graph grows by each new node linking to a number of existing nodes. When a new node v joins the network at time t, it connects randomly to existing nodes (ambassadors $ambs$). Subsequently, for each node $w \in ambs$, the node v performs the following process:

1. Connects to a set of out-neighbors and a set of in-neighbors of the node w. The size of these sets is selected according to a "forward burning" geometric distribution with mean p, and a "backward burning" ratio r.
2. Performs step 1 recursively with the newly connected nodes picked from the out- and in-neighbors.

The Forest Fire model has four parameters: n, the number of required nodes, p, forward burning probability, r, backward burning ratio, and the number of ambassador vertices.

3.2 Barabasi-Albert

Barabasi and Albert [1] proposed an incremental growth model which depends on the preferential attachment mechanism which means that nodes with higher degree have high probability to receive new links, often called "rich get richer"-scheme. In this model, when a new node v joins the network at time t, it connects to existing nodes n depending on a probability p. This probability p is based on the indegree value of the nodes n in order to achieve the concept of the preferential attachment.

The Barabasi-Albert model has two parameters: n, number of required nodes, and m, number of connections performed by each new node with existing nodes.

4 Methodology

Our goal is to propose a new model which generates synthetic graphs that adequately match real retweet-based graphs. For this reason, we need real Twitter datasets and an effective way to evaluate the accuracy of our model in the modeling of real retweet-based graphs.

In this section, we answer the question: "How do we determine that the generated synthetic graphs by the different models are adequately representative of the real retweet-based graphs?" Sala et al. [21] have adopted structure-based graph metrics (statistical similarity metrics) that capture different statistical properties of social graphs. They have used those metrics to figure out the accuracy of synthetic generated graph in capturing statistical properties from friendship-based graphs of Facebook social network. Correspondingly, we evaluate the model's accuracy by measuring the difference between the structural properties of real retweet-based and synthetic graphs. According to [10, 11], the group of structure-based metrics we use to determine the accuracy of the graph models are:

- Node Degree Distribution (NDD) which is the probability distribution of node degree over the whole network. Since the retweet-based graphs are directed, we need for Node Indegree Distribution (NID) and Node Outdegree Distribution (NOD) [3].
- Clustering Coefficient (CC) which is the ratio of the number of links that exist between a node's immediate neighborhood and the maximum number of links that could exist. For the directed graphs, we use Transitivity Coefficient (TC) [29]. With this definition, $0 \leq TC \leq 1$, and $TC = 1$ if the network contains all possible edges (perfect transitivity).
- Diameter (Dia) which is the length of the shortest path between the most distanced nodes.
- Average Path Length (APL) which is the average number of hops along the shortest paths for all possible pairs of nodes.
- Assortativity Coefficient (AC) which measures the level of homophily of the graph [14, 16]. In our study, we use the node degree as an attribute of the homophily. With this definition, $-1 \leq AC \leq 1$, and AC is high if the connected nodes tend to have the same attribute values (degree in our case).

Figure 1 describes graphically our approach in this study. Actually, several strategies have also been proposed to evaluate graph similarity such as [5, 7, 12, 17, 24, 26]. In this study, we have used a statistical strategy because of the computational complexity of the other approaches for large datasets.

As mentioned above, each model is parameterized by different variables. Thus, we face another question: How do we determine the model parameters value to obtain the best synthetic graphs that can acceptably match the real retweet-base graphs? Forest Fire and Barabasi-Albert models have the parameter n (static parameter) which can be determined easily by the number of nodes in the real graphs. In other words, for each real graph, we generate a synthetic counterpart

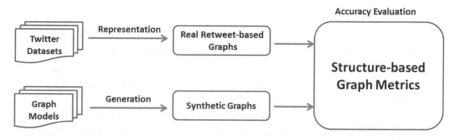

Fig. 1 Our approach to evaluate the accuracy of graph models

with the same nodes number. The Barabasi-Albert model has also the parameter m (static parameter) which detects the number of links performed by each new node to existing nodes. Thus, the total number of links in a generated graph by the Barabasi-Albert model is $n * m$. Accordingly, for each real graph with links E, we generate a synthetic counterpart by setting the parameter m as $m = |E|/n$. The additional parameters of the Forest Fire model (non-static parameters) can be determined by searching the possible parameters value space, and leading the searching using the statistical similarity metrics between the real graph and the model synthetic generated graph. Specifically, we search in the parameters value space for the best values that provide synthetic graphs with the best representative of real graphs according to the statistical graph metrics.

5   Evaluation

In this section, we describe our datasets, and the characteristic of each type. Furthermore, the experimental results on the Forest Fire and the Barabasi models are discussed in this section.

5.1   Datasets

Retweet-based graphs are those graphs in which the links represent the retweet activities between their nodes. Figure 2 shows an example for the retweet-based graph representation where nodes a and d retweeted a tweet of node b whereas the node c retweeted a tweet of the node a.

In this study, we need real Twitter datasets to examine which model generates synthetic graphs that best match the retweet-based graphs which are generated from Twitter datasets. For this reason, we have used three retweet-based datasets. First, we gathered a dataset via Twitter API from December 31, 2015 to January 06, 2016. The collected dataset is the data of hashtag #EndTaizSiege (14,944 nodes and 46,552 connections) that comprises a big connected component (containing 84% of nodes), singletons (14%), and smaller components (2%).

Fig. 2 An example for the
retweet-based graph
representation where nodes a
and d retweeted a tweet of
node b whereas the node c
retweeted a tweet of the
node a

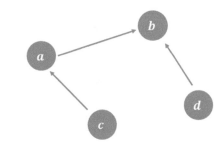

Table 1 Real retweet-based
graphs description

Graph	#Nodes	#Links
#Coup_Suffocates_Taiz (Yemen)	1418	2801
#EndTaizSieg (Yemen)	12,523	32,286
#LaSalida (Venezuela)	22,732	31,179

Second, we collected another dataset from Twitter from July 25 to July 30 in 2016. This Twitter dataset relates to the hashtag #coup_suffocates_Taiz (2241 nodes and 4419 connections) that comprises a big connected component (containing 1418 nodes). The first and second datasets were introduced in [18–20].

The third dataset is the Twitter data of the hashtag #LaSalida with a big connected component containing 22,732 nodes and 31,179 links. All the datasets we used in this study are related to the political uprisings in Yemen and Venezuela. A brief overview of our real retweet-based graphs is shown in Table 1.

5.2 Experimental Results

In this section, we express the results of the models accuracy test using the different structure-based metrics. For the model accuracy test, we generate 40 synthetic graphs from Forest Fire and Barabasi-Albert models for each real retweet-based graph, and compute the difference between the average value of the 40 synthetic graphs and the real graph counterpart. As mentioned above, we have to determine the parameter values to produce the best synthetic graphs that can acceptably match the real retweet-base graphs. According to our approach, we found that the best values of the non-static parameters of the Forest Fire model are: $p = 0.35, r = 0.30$, and $ambs = 1$.

Related to the node degree distribution metric, we evaluate the results using the Kolmogorov-Smirnov test [22, 28]. The Kolmogorov-Smirnov test returns a $D \in [0, 1]$ statistic and a p-value corresponding to the D statistic. The D statistic is the absolute max distance between the probability distribution of the two samples. The closer this number is to 0 the more likely it is that the two samples were drawn from the same distribution.

Table 2 Kolmogorov-Smirnov test between synthetic graphs (derived from forest fire and Barabasi-Albert models) and the real graphs for degree distribution metric

Graph	Model	Kolmogorov-Smirnov test (D value) : real vs. synthetic graphs	
		NID	NOD
Taiz/1416 nodes	Forest fire	0.274	0.387
	Barabasi-Albert	0.212	0.916
Taiz/12522 nodes	Forest fire	0.185	0.378
	Barabasi-Albert	0.039	0.933
Venezuela	Forest fire	0.085	0.316
	Barabasi-Albert	0.108	0.869

Each value is the average of D statistic of 40 synthetic to real graph comparisons. The D statistic values are significant with $\alpha = 0.01$ for NID and NOD

Table 2 shows the max distance D between the probability distribution derived from the real and synthetic graphs. From the table, the D statistic values are almost similar (close to 0) in case of the node indegree distribution for both models. Nevertheless, with respect to the node outdegree distribution, the Forest Fire model outperforms the Barabasi-Albert model. Thus, we conclude that:

- Forest Fire model produces relatively accurate representation of the real graph with respect to the node degree distribution metric.
- Barabasi-Albert model does not success to represent the real graphs of all real Twitter datasets with respect to the node outdegree distribution metric. This is reasonable since the outdegree values of the generated graph nodes are identical, and they are equal to m. However, it produces relatively accurate representation of the real graph with respect to the node indegree distribution metric.

Table 3 shows the application of the additional structure-based metrics on generated synthetic graphs by each model and the real retweet-based graphs. From the table, we got the following outline:

- The Barabasi-Albert model is relatively accurate for the transitivity metric. Whereas, it generates synthetic graphs whose transitivity values approximate the transitivity values of the real graphs with respect to all real datasets.
- As well as, the Forest fire model is relatively accurate with respect to the average path length.
- Both models produce relatively accurate representation of the real graphs with respect to the diameter metric.
- Barbasi model outperforms Forest Fire model with respect to the assortativity and the transitivity metrics.

We observe that the Forest Fire model generates synthetic graphs with high transitivity compared to the real graphs. This happens because of the links density which is produced by each new incoming node to existing nodes. To decrease the density, we can decrease the burning probability p and r. According to the

Table 3 Transitivity, diameter, average path length, and assortativity of synthetic graphs (derived from forest fire and Barabasi-Albert models) and the real graphs. Results exposed are the exact metric value, and results for synthetic graphs are averages of 40 graphs

Graph	Metrics value			
	TC	D	APL	AS
Taiz/1416 nodes	0.025	10	4.06	−0.27
Forest fire	0.25	11.3	3.403	0.271
Barabasi-Albert	0.007	9.3	2.7	−0.137
Taiz/12522 nodes	0.025	16	5.14	−0.251
Forest fire	0.15	14.5	4.487	0.115
Barabasi-Albert	0.002	12.55	3.432	−0.122
Venezuela	0.0003	18	6.03	−0.246
Forest fire	0.121	16.4	4.65	0.11
Barabasi-Albert	0	15.2	3.613	−0.043

nature of Forest Fire model, decreasing the burning probability leads to decrease the transitivity, but the node degree distribution and diameter metrics will be effected. To solve this problem, we decreased the burning probabilities and increased the number of ambassadors $ambs$. We found also that the density of generated graphs increases, and this leads to decrease the diameter and increase the transitivity as well. The decreasing burning probability or decreasing the number of ambassadors in the Forest Fire model unconditionally effects on the number of links which performed by each new incoming node to the existing nodes (just depending on burning probability). We need a way through which we can decrease the links density of each node according to some conditions.

In the next section, we describe our proposal to extend the Forest Fire model with some modifications to enhance the model performance to convincingly represent real retweet-based graphs. In the proposed model, we increase the number of ambassadors to be 2, and decrease the generated graphs density by adding some conditions in the burning process (not by decreasing the burning probabilities p and r). Thus, the burning probabilities in the extend Forest Fire model will stay as in the original one.

6 Extended Forest Fire

The proposed extended Forest Fire is a hybrid model that acts like Forest Fire model with a modification adapted from Barabasi-Albert model. Specifically, we adopt the preferential attachment through the burning process of the Forest Fire model.

In our extended Forest Fire model, when a new node v joins the network at time t, it connects randomly to existing nodes $ambs$. Subsequently, for each node $w \in ambs$, the node v performs the following process:

1. It selects existing out-neighbors n and in-neighbors m of the node w. The size of the nodes n and m is respectively selected according to a forward burning probability p and a backward burning ratio r.

2. It connects to the existing nodes n and m. The probability that an existing node $v_i \in (n \cup m)$ is chosen is given by:

$$bp(v_i) = \frac{1 + \text{indegree}(v_i)}{1 + \text{max(indegree)}} \qquad (1)$$

where max(indegree) is the maximum indegree value in the graph.

3. It performs steps 1 and 2 recursively with newly connected nodes n and m.

To determine the power of the preferential attachment process, we insert a new parameter γ $(0 \leq \gamma \leq 1)$ in formula (1). Thus, the probability that the existing node v_i is chosen is given by:

$$bp(v_i) = \left(\frac{1 + \text{indegree}(v_i)}{1 + \text{max(indegree)}} \right)^{\gamma} \qquad (2)$$

As in the determination of the best values of the burning probabilities in Forest Fire model, we determine the best value of the parameter γ that provides synthetic graphs with the best representative of real graphs according to the statistical graph metrics. Figure 3 shows how the extended Forest Fire generates synthetic graphs in which the node degree follows the power-law distribution.

6.1 Experimental Results

Using the structure-based metrics, we evaluate the accuracy of our extended Forest Fire model comparing to the Forest Fire and Barabasi-Albert models. Table 4 shows the performance of the extended Forest Fire model with respect to the degree distribution metric. As above, we can interpret the D statistic values to conclude that our extended model provides highly accurate representation of the real retweet-based graphs regarding the node indegree distribution (D is very close to 0). Furthermore, it produces relatively accuracy related to the node outdegree distribution. In both cases, it outperforms the Forest Fire and the Barabasi-Albert models (Fig. 4).

Table 5 expresses the performance of the extended Forest Fire model with respect to the other metrics. We notice that the extended Forest Fire model:

– provides sensibly enhancement over the Forest Fire and the Barabasi-Albert models with respect to the average path length metric.
– provides sensibly enhancement over the Forest Fire model with respect to the transitivity and the assortativity metrics.
– has also relatively accurate representation of the real graphs with respect to the diameter metric.

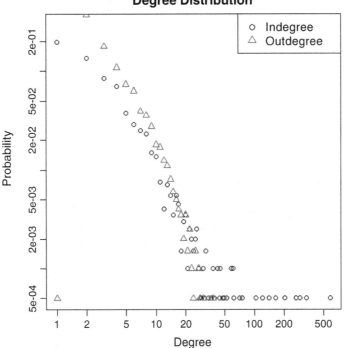

Fig. 3 Degree distribution of generated graph by the extended Forest Fire model (number of nodes = 2000, forward burning probability: 0.40, backward burning probability: 0.35, and attachment power: $\gamma = 0.25$)

Table 4 Kolmogorov-Smirnov test between synthetic graphs (derived from extended forest fire, fire forest, and Barabasi-Albert models) and the real graphs for degree distribution metric. Each value is the average of D statistic of 40 synthetic to real graph comparisons. The D statistic values are significant with $\alpha = 0.01$ for NID and NOD

Graph	Model	Kolmogorov-Smirnov test (D value) : real vs. synthetic graphs	
		NID	NOD
Taiz/1416 nodes	Extended forest fire	0.131	0.272
	Forest fire	0.274	0.387
	Barabasi-Albert	0.212	0.916
Taiz/12522 nodes	Extended forest fire	0.025	0.362
	Forest fire	0.185	0.378
	Barabasi-Albert	0.039	0.933
Venezuela	Extended forest fire	0.061	0.322
	Forest fire	0.085	0.316
	Barabasi-Albert	0.108	0.869

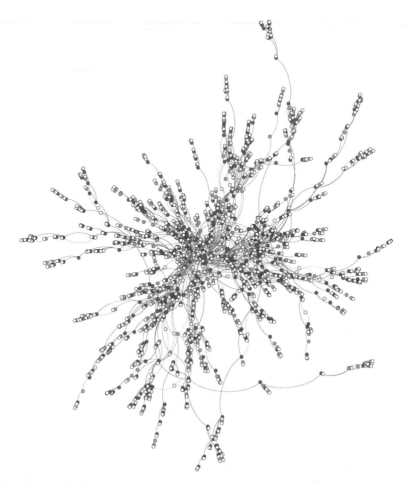

Fig. 4 An example for a synthetic graph generated by the extended Forest Fire model (number of nodes = 2000, forward burning probability: 0.40, backward burning probability: 0.35, and attachment power: $\gamma = 0.25$)

Briefly, depending on the preferential attachment during the burning process of the Forest Fire model, the extended Forest Fire provides the most accurate synthetic graphs that represent the real graphs comparing to the feature-driven models, Forest Fire and Barabasi-Albert models. Specifically, it outperforms the pre-proposed models in capturing the statistical similarity metrics of the real retweet-based graphs.

Table 5 Transitivity, diameter, average path length, and assortativity of synthetic graphs (derived from extended forest fire, forest fire, and Barabasi-Albert models) and the real graphs. Results exposed are the exact metric value, and results for synthetic graphs are averages of 40 graphs

Graph	Metrics value			
	TC	D	APL	AS
Taiz/1416 nodes	0.025	10	4.069	−0.27
Extended forest fire	0.041	13.45	4.055	−0.083
Forest fire	0.25	11.3	3.403	0.271
Barabasi-Albert	0.007	9.3	2.7	−0.137
Taiz/12522 nodes	0.025	16	5.14	−0.251
Extended forest fire	0.032	19.4	5.14	−0.073
Forest fire	0.15	14.5	4.487	0.115
Barabasi-Albert	0.002	12.55	3.432	−0.122
Venezuela	0.0003	18	6.03	−0.246
Extended forest fire	0.009	20.1	5.91	−0.073
Forest fire	0.121	16.4	4.65	0.11
Barabasi-Albert	0	15.2	3.613	−0.043

7 Conclusion

The goal of this article was to derive a model for Twitter retweet networks—in particular for the proposed datasets based on political uprisings. We identified feature-driven models as the type of models which best fit this purpose. To evaluate the quality of fit of a model we introduced graph properties and evaluated two existing models (Barabasi-Albert, Forest Fire). Next, to improve the fit we proposed a new model combining the Forest Fire model with an acceptance probability known from Barabasi-Albert. Our final evaluation shows an improvement of fit in particular for node degree distribution, transitive closure, and average path length. While the combined model, involving two distinct steps, is slightly more complex than the classical models, the additional acceptance parameter allows more precise fine-tuning. Further studies could evaluate the applicability of the model to different use cases.

References

1. Barabsi, A.L., Albert, R.: Emergence of scaling in random networks. Science **286**(5439), 509–512 (1999)
2. Barrett, C.L., Beckman, R.J., Khan, M.K., Kumar, V.S.A., Marathe, M.V., Stretz, P.E., Dutta, T., Lewis, B.: Generation and analysis of large synthetic social contact networks. In: Proceedings of the Winter Simulation Conference (WSC), pp. 1003–1014. IEEE, New York (2009)
3. Clauset, A., Shalizi, C.R., Newman, M.E.J.: Power-law distributions in empirical data. SIAM Rev. **51**(4), 661–703 (2009)
4. Erdos, P., Rényi, A.: On random graphs, I. Publ. Math. (Debrecen) **6**, 290–297 (1959)
5. Fernández, M.L., Valiente, G.: A graph distance metric combining maximum common subgraph and minimum common supergraph. Pattern Recogn. Lett. **22**(6), 753–758 (2001)

6. Gilbert, E.N.: Random graphs. Ann. Math. Stat. **30**(4), 1141–1144 (1959)
7. Leicht, E.A., Holme, P., Newman, M.E.J.: Vertex similarity in networks. Phys. Rev. E **73**(2), 026120 (2006)
8. Leskovec, J., Faloutsos, C.: Scalable modeling of real graphs using Kronecker multiplication. In: Proceedings of the 24th International Conference on Machine Learning, pp. 497–504. ACM, New York (2007)
9. Leskovec, J., Kleinberg, J., Faloutsos, C.: Graphs over time: densification laws, shrinking diameters and possible explanations. In: Proceedings of the Eleventh ACM SIGKDD International Conference on Knowledge Discovery in Data Mining, pp. 177–187. ACM, New York (2005)
10. Mahadevan, P., Krioukov, D., Fall, K., Vahdat, A.: Systematic topology analysis and generation using degree correlations. ACM SIGCOMM Comput. Commun. Rev. **36**(4), 135–146 (2006)
11. Mahadevan, P., Krioukov, D., Fomenkov, M.F., Dimitropoulos, X., Vahdat, A.: The internet as-level topology: three data sources and one definitive metric. ACM SIGCOMM Comput. Commun. Rev. **36**(1), 17–26 (2006)
12. Messmer, B.T., Bunke, H.: A new algorithm for error-tolerant subgraph isomorphism detection. IEEE Trans. Pattern Anal. Mach. Intell. **20**(5), 493–504 (1998)
13. Mitzenmacher, M.: A brief history of generative models for power law and lognormal distributions. Internet Math. **1**(2), 226–251 (2004)
14. Newman, M.E.J.: Assortative mixing in networks. Phys. Rev. Lett. **89**(20), 208701 (2002)
15. Newman, M.E.: The structure and function of complex networks. SIAM Rev. **45**(2), 167–256 (2003)
16. Newman, M.: Networks: An Introduction, pp.1–2. Oxford University Press, New York (2010)
17. Park, K., Han, Y., Lee, Y.K.: An efficient method for computing similarity between frequent subgraphs. In: Third International Conference Cloud and Green Computing (CGC), pp. 566–567. IEEE, New York (2013)
18. Qasem, Z., Jansen, M., Hecking, T., Hoppe, H.U.: Detection of strong attractors in social media networks. Comput. Soc. Netw. **3**(1), 11 (2016)
19. Qasem, Z., Jansen, M., Hecking, T., Hoppe, H.U.: Influential actors detection using attractiveness model in social media networks. In: International Workshop on Complex Networks and their Applications, pp. 123–134. Springer, Berlin (2016)
20. Qasem, Z., Jansen, M., Hecking, T., Hoppe, H.U.: Using attractiveness model for actors ranking in social media networks. Comput. Soc. Netw. **4**(1), 3 (2017)
21. Sala, A., Cao, L., Wilson, C., Zablit, R., Zheng, H., Zhao, B.Y.: Measurement-calibrated graph models for social network experiments. In: Proceedings of the 19th International Conference on World Wide Web, pp. 861–870. ACM, New York (2010)
22. Smirnov, N.: Table for estimating the goodness of fit of empirical distributions. Ann. Math. Stat. **19**(2), 279–281 (1948)
23. Solomonoff, R., Rapoport, A.: Connectivity of random nets. Bull. Math. Biol. **13**(2), 107–117 (1951)
24. Song, C., Havlin, S., Makse, H.A.: Self-similarity of complex networks. Nature **433**(7024), 392–395 (2005)
25. Tan, V.Y.F., Sanghavi, S., Fisher, J.W., Willsky, A.S.: Learning graphical models for hypothesis testing and classification. IEEE Trans. Signal Process. **58**(11), 5481–5495 (2010)
26. Topirceanu, A., Udrescu, M.: Statistical fidelity: a tool to quantify the similarity between multivariable entities with application in complex networks. Int. J. Comput. Math. **94**, 1–19 (2016)
27. Vazquez, A.: Growing network with local rules: preferential attachment, clustering hierarchy, and degree correlations. Phys. Rev. E **67**(5), 056104 (2003)
28. Wang, J., Tsang, W.W., Marsaglia, G.: Evaluating Kolmogorov's distribution. J. Stat. Softw. **8**, 18 (2003)
29. Watts, D.J., Strogatz, S.H.: Collective dynamics of small-world networks. Nature **393**(6684), 440–442 (1998)
30. Zweig, K.A.: Network Analysis Literacy: A Practical Approach to the Analysis of Networks. Springer, Berlin (2016)

Incremental Learning in Dynamic Networks for Node Classification

Tomasz Kajdanowicz, Kamil Tagowski, Maciej Falkiewicz, and Przemyslaw Kazienko

Abstract An incremental learning method for nodes' classification is presented in the paper. In particular, there is proposed an active scheme algorithm for multi-class classification of nodes' states that varies over time and depends on information spread in the network. Demonstration of the method is conducted using social network dataset. According to sent messages between nodes, the emotional state of the message sender updates each receiving node's feature vector and the method tries to classify next emotional state of the receiver. The novelty of the proposed approach lies in applying incremental learning method for non-stationary network environment. There are demonstrated some properties of the proposed method in experiments with real data set, showing that the method can effectively classify the future state of nodes.

Keywords Relational classification · Incremental learning · Classification · Emotional classification · Social networks

1 Introduction

In this paper, we describe and analyze an online learning task in a network environment for the purpose of node classification. We first introduce a simple scheme for online learning that is feasible to handle dynamic and even multi-layer networks. Then the real-world scenario for classification of the users' mood in an online virtual world platform—Timik.pl—available for users in Poland was tested. Based on the analysis of emoticons extracted from chat messages sent between

T. Kajdanowicz (✉) · K. Tagowski · M. Falkiewicz · P. Kazienko
Department of Computational Intelligence, Wroclaw University of Science and Technology, Wroclaw, Poland
e-mail: tomasz.kajdanowicz@pwr.edu.pl

© Springer International Publishing AG, part of Springer Nature 2018   133
R. Alhajj et al. (eds.), *Network Intelligence Meets User Centered Social Media Networks*, Lecture Notes in Social Networks,
https://doi.org/10.1007/978-3-319-90312-5_9

users, the incremental learning method was able to train classifiers and through them uncover proper moments of mood change as well as resulting new mood. It was obtained by capturing all the interaction between nodes in the online data stream.

The main focus of the paper is on incremental, a.k.a. online, learning within a network. In the standard incremental setting, a learning algorithm observes data cases in a sequential, streaming manner. The classification outcome is predicted after each data case arrives. The outcome can be simple (e.g., 0/1) as in binary classification or have a form of more complex structure (e.g., multi-class, multi-label, structured output). An indispensable feature of the incremental learning methods, once the algorithm made a prediction, is a loop that provides feedback indicating the correct outcome. Based on it the algorithm can modify its classification mechanism rising the chances for correct classification of upcoming data cases. In general, the construction of incremental learning algorithms and methods is straightforward and simple to implement, but in case of network data might provide wide bounds on their performance.

The main motivation to consider the emotional state classification within the incremental learning setting is twofold. Firstly, with the development of the internet, every user has received an opportunity to express their opinion and emotional state. Because current state of the digital world can be characterized as a continuous interaction between internet users, the obvious observation is that any such interactions can cause dependencies between users. The emotional state of one user may easily affect his all interlocutors. We would like to propose a method for quantifying such phenomenon. Secondly, the aforementioned relations and dependencies of emotional states in social networks are very dynamic and may have various underlying phenomena in time. To address such problem, an incremental learning scheme is considered.

The paper is a short introduction to preliminary results of our research and is organized as follows: the concise presentation of related work in the related fields of classification in networks as well as incremental learning is presented in Sect. 2. In Sect. 3 there is described the considered problem, as well as a new incremental learning method for node classification is proposed. The experimental results achieved on real-world data with evaluation of the algorithms' accuracy are gathered in Sect. 4 and concluded in Sect. 5.

2   Related Work

2.1   *Classification in Networks*

One of the most important directions in research on networks is node classification. Classification of nodes in the network may be performed either by means of known profiles of these nodes (the regular concept of classification) or based on information derived from the interconnection of nodes in the network—so-called collective classification [12]. An example of such problem is web page categorization based on categories of pages connected to it. It is very likely that a given web page is related to the sport if it is linked by many other web pages about the sport.

There exist a variety of methods for collective classification, e.g. [12, 13]. However, there can be distinguished two types of them: local and global. The former methods use classifiers trained on nodes' features obtained from their structural features whereas the latter are defined as a pre-defined process with global objective function.

Additionally, classification of nodes in the network can be solved using two distinct approaches: within-network and across-network inference. Within-network classification [1], for which training nodes are connected directly to other nodes, whose labels are to be classified, stays in contrast to across-network classification [9], where models learned from one network are applied to another similar network.

There are related several problems with a collective classification that have been currently addressed by researchers. One of them is the problem of what features should be used to maximize the classification accuracy. In approaches which use local classifiers, the relational domain in the form of networks' topology needs to be transformed to vector notation by calculating proper characteristics for each node. It has been reported that precise solution strongly depends on the application domain [11]. The previous research showed that new attribute values derived from the graph structure of the network, such as the betweenness centrality, may be beneficial to the accuracy of the classification task [3]. It was also confirmed by other research discussed in [7].

Another interesting problem in collective classification based on iterative algorithms is the ordering strategy that determines in which order to visit the nodes iteratively to re-label them. The order of visiting the nodes influences the values of input features that are derived from the structure. A variety of sophisticated or very simple algorithms can be used for this purpose. Random ordering that is one of the simplest ordering strategies used with iterative classification algorithms can be quite robust [8].

Two of the most popular local collective classification methods are: Iterative Classification Algorithm (ICA) and Gibbs Sampling Algorithm (GSA), introduced by Geman & Geman in the image processing context [5]. Both of them belong to the so-called approximate local inference algorithms basing on local conditional classifiers [12]. Another technique is a Loopy Belief Propagation (LBP) [10] that is the global approximate inference method used for collective classification.

The proposed new incremental classification method is based on local conditional classifiers that operate in the input space formed by relational features for each node, i.e. vectors denoting emotional states of all incoming communication in egocentric networks. In general, utilized features can be of any kind that represents relational nature of connections in the network, e.g. label-dependent ones [7].

2.2 Incremental Learning

The major trend of data science and in particular of machine learning research has been focused on algorithms that can learn using training data characterized by the fixed, unknown distribution. The assumption of the stationary distribution was even applied in online and incremental learning, i.e. various forms of neural networks and other statistical approaches while performing partial fitting are based on the idea that data distribution is constant [6, 14]. The problem of learning in changeable, non-stationary environment, where the distribution of data changes over time, has received less attention. Especially when it was devoted to classification in a relational domain—for the purpose of node classification. As more and more practical problems required sophisticated learning techniques, the non-stationary environment based learning methods have received increasing attention [2]. It is getting even more complicated while the learning should be able to consume the stream of interaction and relations of an interconnected node in the graph. Hence, the incremental learning in dynamic networks for node classification is closely related to the field of game theory, information theory, especially information diffusion influence spread as well as recently machine learning.

There can be distinguished two primary families of common approaches used for incremental learning. They are referred to as *active* and *passive* strategies [2]. The main difference between them lays in the adaptation mechanism to address the change in the data stream: active approaches explicitly detect the change in the data distribution and only then adapt algorithm through some update mechanism, whereas passive approaches update the model continuously in time without requiring an explicit detection of the change.

3 Incremental Learning Method for Node Classification in Network

3.1 Problem Description

Assuming that $D_t(y)$ is the distribution providing the prior probability of a single node changing its class into one of the discrete classes $y \in \Lambda$ that constitutes the communication in the network let us characterize the incremental learning problem in the network domain. This distribution is sub-scripted with time t to explicitly show its time-varying nature. Let Y be a random variable having distribution $D_t(y)$ and $P(Y \mid (X, c))$ be the posterior probability of such communications, where $X \in R^d$ represents a feature vector modelled as a random variable describing all incoming communication to a particular node, and c is the current nodes class. The incremental learning problem in networks is then addressing a situation when data arrive in an online manner, i.e., one single instance at a time, or in a batch setting, and appropriately, the classifier Φ is being trained (its Θ configuration is obtained) using such data.

In particular, the proposed method addresses the case of batch setting that provided a finite set of tuples $S_t = \{(X^1, c^1), \ldots (X^N, c^N)\}$ perform additional training of classifier: $\theta_{t+1} \leftarrow additional\_training(\Phi, \theta_t, S_t)$. It is proposed to perform such additional training whenever some quality measure is dropped for consecutive batches. It comes from a straightforward philosophy that as long the classifier generalizes appropriately recent data it should not be additionally trained. However, in case of decrease of classification quality on the current batch, it should be additionally trained. A single batch can contain fixed number of tuples or tuples from a fixed time period.

In general, incremental learning tries to capture changes in the data that arise in time. There may be distinguished two causes or natures of changes in the data: *real drift*, where posterior probability $P(Y \mid (x, c))$ varies over time, independently from variations in the prior probability $D_t(y)$ and *virtual drift*, where the evidence or the marginal distribution of the data, $P(X)$, changes without affecting the posterior probability of classes $P(Y \mid (X, c))$ [4]. Additionally, any change in probability distributions can be characterized by the rate at which the drift is taking place. The change can be abrupt, resulting in a sudden drift, or be gradual, which is defined as slowly evolving distributions. Due to the fact that identification of the type of change in the distribution can be difficult, the common technique is to observe the quality of classification and to perform additional incremental learning when the result worsen.

3.2 The Incremental Iterative Classification Algorithm (IICA)

In order to capture the changes in distributions in the network, we introduce a new Incremental Iterative Classification Algorithm (IICA). It provides four important parts of incremental learning process, i.e. extraction of features from network events (e.g., textual communication (m) between pair of nodes (v_i, v_j)), classification part that provides classes for network nodes', change detection functionality, that observes the results of classification and when it is worsening launches last part—classifier adaptation mechanism, that incrementally trains a classifier, please see Fig. 1.

The IICA algorithm requires a classifier Φ with its initially estimated parameters Θ, please see Algorithm 1. This means that the obvious standard procedure of training on the initial data might be required before launching IICA. The proposed method is consuming batches with a fixed number of events (S_t) that consists of communication events between a pair of nodes, and a message sent (v_j, v_i, m). Using such events each receiving node accumulates the knowledge of the state of the sender at the time the event occurred. This may be derived directly from senders' properties or the transferred message. Using all of the communication events in batch, see Algorithm 2, the method updates the feature map of the receiver, here proposed simple sum aggregation. Then all receivers from mini-batch are classified using newly obtained values of features, and the results of classification

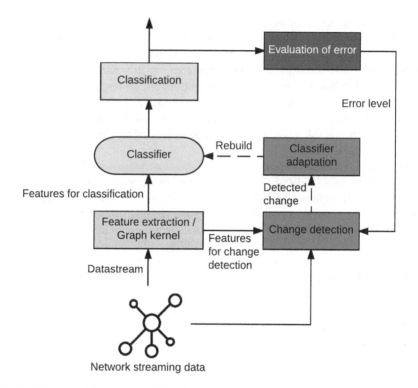

Fig. 1 Main steps of the incremental learning approach of a classifier in non-stationary environment using network streams

---

Algorithm 1 Incremental Iterative Classification Algorithm (IICA)

Ensure: classifier Φ with initial Θ parameters
1: results = []
2: train_data = []
3: **for** $mini\_batch \in stream$ **do**
4:     update_features($mini\_batch$)
5:     classes $\leftarrow \Phi(mini\_batch, \Theta)$
6:     results.append(evaluation($classes,mini\_batch$))
7:     **if** $change\_detect(result[:-1], result[:-2]))$ **then**
8:         $\theta \leftarrow additional\_training(\Phi, \theta, train\_data)$
9:         train_data = []
10:    **end if**
11:    train_data += mini_batch
12: **end for**

---

are evaluated and stored in result history. It might be debatable that the evaluation of the classifier performance is impossible as long as there are no collected true classes from the network. Indeed in such case, the procedure is suspended until the real data arrives and all other mechanisms that would be able to deal with such problem are not discussed here. Then the method applies simple mechanism described below that tries to capture whether there should be performed additional training on new data. For that purpose change detection ratio (cdr) is calculated, see Algorithm 3, from quality results of classification in current and previous batch. Whenever the ratio is smaller than the threshold, the classifier Φ is additionally trained on the data that arrived from the time of its previous incremental learning.

Algorithm 2 The pseudo-code for *update_features* subroutine

Ensure: NF - nodes feature map, v_j - receiving node, v_i - sending node, m - message
1: **for** each *event* $\in$ *mini_batch* **do**
2:     **if not** NF.contains(v_j) **then**
3:         $NF(v_j)$ = initialise_features()
4:     **end if**
5:     $NF(v_j).values$ += features(v_i, m)
6: **end for**

Algorithm 3 The pseudo-code for *change_detect* subroutine

Ensure: *threshold* - additional train ratio, *curr_result* - current evaluation of results, *prev_result* - previous evaluation of results
1: $cdr = \frac{curr\_result - prev\_result}{curr\_result}$
2: **if** $cdr <$ *threshold* **then**
3:     **return** true
4: **else**
5:     **return** false
6: **end if**

---

4 Experiments and Results

In order to evaluate the preliminary proposal of the incremental classification method in the network domain, it was tested on the real-world dataset in the task of emotional state classification.

4.1 Dataset

The dataset comes from the system that enabled users, represented by graphical avatars, to express their current state and gave the opportunity to engage in the life of an online community. Since the system launch in 2007, users have activated more than 850,000 accounts. The platform was fully operational in years 2010–2012. The used dataset is composed of 16M chat messages between pairs of nodes as well as

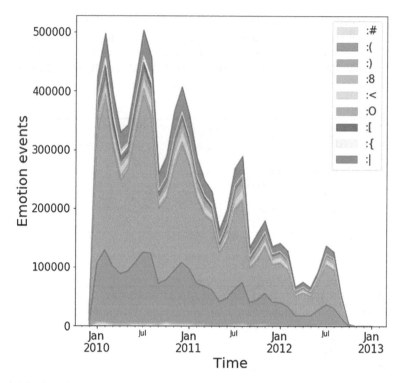

Fig. 2 Number of messages with emotional expressions by type of emotion in lifetime of the *Timic* social network platform

of the emotional state of these nodes. Moreover for each of the nodes all emotional state changes are implicitly given—nine emotions are quantified by emoticons ':#',':(',':)',':8',':<',':O',':[',':{',':|' that appear in the chat discussions and are used in the status of avatars. The emotions mentioned above are taken as nodes' class labels. Their semantics, as well as all relation to well-known dichotomies of emotions, is out of the scope of the paper. The number of users having a particular emotional state in time is given in Fig. 2.

4.2 Experimental Protocol and Results

The experiments cover the situation when there is known the moment of the emotional switch (update or change) for nodes and our task is to predict the new emotional state. This translates the problem to the classification of emotional states as a multi-class task. In order to learn the classifiers for emotional change there were build features for each node that contained the number of interactions with all neighbors broken down into each individual emotions. Additionally, the feature map was extended with information of previous emotional state. Thus, all together there were 10 features used for generalization.

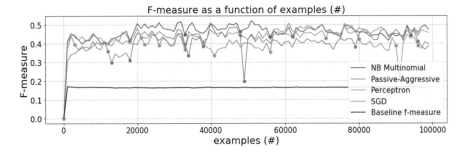

Fig. 3 F-measure classification results of 4 classifiers in the incremental learning approach for graph data stream with a base-line f-measure. The dots denote a point in time when change was detected and the classifier was additionally trained (first 100k emotion changes)

There were employed four classifiers that are capable of incremental learning: SGD (SVM with stochastic gradient descent learning), Perceptron, Multinomial Naive Bayes, and Passive-Aggressive. Such classifiers were selected knowing their various characteristics, that should not affect the overall applicability of the proposed framework and confirm its robustness. In order to evaluate the classifier quality weighted f-measure was used and the threshold for change detection ratio was set to -0.1. Evaluation of current states of classifiers (initially and additionally trained) was performed on batches equal to 1000 examples that appeared in the next window.

According to the first 100k results presented in Fig. 3 the IICA method demonstrates its ability to recover from worsen classification results in time. It is clearly visible for data cases just after 60,000 examples, where spectacular drop of f-measure was observed for Passive-Aggressive and Perceptron algorithms. After additional training rounds (dots on Figure) the algorithm performs much better. The overall classification quality is reasonable and is much better than baseline f-measure (if the result was randomly drawn).

5   Conclusions and Further Work

An incremental learning method for nodes' classification was presented in the paper. Its abilities to classify nodes were demonstrated with the task of nodes' emotional state categorization from communication in the social network. The obtained results were promising, and further research will be focused on more comprehensive change detection mechanisms as well as an adaptation of deep neural networks in incremental learning in networks.

Acknowledgements The work was partly supported by The Polish National Science Centre, project no. 2013/09/B/ST6/02317 and 2016/21/D/ST6/02948 as well as the European Union's Horizon 2020 research and innovation programme under the Marie Sklodowska-Curie grant agreement No 691152, RENOIR project; the Polish Ministry of Science and Higher Education fund for supporting internationally co-financed projects in 2016–2019 (agreement no. 3628/H2020/2016/2). The calculations were carried out in the Wroclaw Centre for Networking and Supercomputing[1], grant No 177.

References

1. Desrosiers, C., Karypis, G.: Within-network classification using local structure similarity. In: Machine Learning and Knowledge Discovery in Databases. Lecture Notes in Computer Science, vol. 5781, pp. 260–275. Morgan Kaufmann, San Francisco (2009)
2. Elwell, R., Polikar, R.: Incremental learning of concept drift in nonstationary environments. IEEE Trans. Neural Netw. **22**(10), 1517–31 (2011). A publication of the IEEE Neural Networks Council
3. Gallagher, B., Eliassi-Rad, T.: Leveraging label-independent features for classification in sparsely labeled networks: an empirical study. In: Proceedings of Second ACM SIGKDD Workshop on Social Network Mining and Analysis, SNA-KDD'08 (2008)
4. Gama, J., Žliobaitė, I., Bifet, A., Pechenizkiy, M., Bouchachia, A.: A survey on concept drift adaptation. Comput. Surv. **46**(4), 1–37 (2014)
5. Geman, S., Geman, D.: Stochastic relaxation, Gibbs distributions and the Bayesian restoration of images. IEEE Trans. Pattern Anal. Mach. Intell. **6**, 721–741 (1984)
6. Hand, D.J.: Classifier technology and the illusion of progress. Stat. Sci. **21**(1), 1–15 (2006)
7. Kazienko, P., Kajdanowicz, T.: Label-dependent node classification in the network. Neurocomputing **75**(1), 199–209 (2012)
8. Knobbe, A., deHaas, M., Siebes, A.: Propositionalisation and aggregates. In: Proceedings of Fifth European Conference on Principles of Data Mining and Knowledge Discovery, pp. 277–288 (2001)
9. Lu, Q., Getoor, L.: Link-based classification. In: Proceedings of 20th International Conference on Machine Learning (ICML), San Francisco, pp. 496–503 (2003)
10. Pearl, J.: Probabilistic Reasoning in Intelligent Systems. Morgan Kaufmann, San Francisco (1988)
11. Perlich, C., Provost, F.: Distribution-based aggregation for relational learning with identifier attributes. Mach. Learn. **62**(1–2), 65–105 (2006)
12. Sen, P., Namata, G., Bilgic, M., Getoor, L., Gallagher, B., Eliassi-Rad, T.: Collective classification in network data. Artif. Intell. Mag. **29**(3), 93–106 (2008)
13. Taskar, B., Segal, E., Koller, D.: Probabilistic clustering in relational data. In: Seventeenth International Joint Conference on Artificial Intelligence (IJCAI-01), pp. 870–887 (2001)
14. Zliobaite, I., Bifet, A., Pfahringer, B., Holmes, G.: Active learning with drifting streaming data. IEEE Trans. Neural Netw. Learn. Syst. **25**(1), 27–39 (2014)

---

[1] http://www.wcss.wroc.pl.

Sponge Walker: Community Detection in Large Directed Social Networks Using Local Structures and Random Walks

Omar Jaafor and Babiga Birregah

Abstract Community detection is one of the most central topics in social network analysis. Yet, many methods that account for directed interactions in social networks scale poorly. Some methods that have lower complexity focus on local criteria, ignoring larger structures. This paper proposes an iterative community detection algorithm that captures local structure in its first iterations and then performs random walks to capture larger structures on a reduced network. The proposed algorithm returns a dendrogram and also suggests the best partition of a network. Experimentations on synthetic social networks were performed and comparisons with benchmark community detection algorithms show that the Sponge walker detects high quality clusters despite low time complexity.

Keywords Community detection · Social networks · Directed graphs · Big data

1 Introduction

As noted by Malliaros and Vazirgiannis [1], social networks are very different from networks obtained using random graph generators such as Erdo-Renyi [2]. Real networks are characterized by many properties like a degree power law distribution, small diameter, non reciprocal interactions, and dense interactions inside small groups of nodes. Community detection is concerned by the last property of real networks, which is the partitioning of a graph into non-overlapping groups.

Community detection in graphs has focused on undirected graphs, where an edge represents a two-way relationship or interaction. Despite the availability of data from directed networks, community detection in these networks has not

O. Jaafor (✉) · B. Birregah
Charles Delaunay Institute, UMR CNRS 6281, University of Technology of Troyes,
Troyes, France
e-mail: omar.jaafor@utt.fr; babiga.birregah@utt.fr

© Springer International Publishing AG, part of Springer Nature 2018   143
R. Alhajj et al. (eds.), *Network Intelligence Meets User Centered Social Media
Networks*, Lecture Notes in Social Networks,
https://doi.org/10.1007/978-3-319-90312-5_10

received significant attention [1]. In fact, directed networks are often represented by undirected graphs, which leads to ignoring important features of the network.

One of the reasons the undirected graph representation of directed networks is popular is that clustering directed graphs is an inherently difficult task. Fortunato et al. [3] note that the asymmetric representation of graphs makes spectral analysis much more complicated. Few techniques can be extended from undirected graphs [3] leading to the development of new methods which are often greedier.

Another difficulty linked to clustering directed graphs is that some of the core properties of graphs are not well defined [1] (e.g., density). These properties are central to many graph clustering methods that optimize specific measures. Another difficulty is the lack of an agreed upon definition of a cluster in these graphs. The presence of flows in the network implies that there are patterns not easily captured by a measure such as density.

There are many definitions of a cluster in a directed graph that are highly dependent on the application domain where it is used. The most common definition comes from the undirected graph clustering problem. A cluster is defined as a set of nodes with high density intra-cluster edge density and lower inter-cluster edge density. Many quality measures for graph partitions such as modularity [4] were developed with this idea in mind. Recent research works translated the concept of density to directed graphs and proposed new quality measures that account for edge directionality. Using only density based measures leads a clustering algorithm to focus only on local features, ignoring many important patterns that can only be captured by analyzing paths in the graph.

Other definitions of clusters in a directed graph are linked to patterns that emerge from the directionality of edges. Two tendencies were identified:

– Capturing the flow of information in a network: a cluster is a subnetwork that traps a random walker
– Capturing the similarities of interactions: The nodes that compose a cluster link to the same set of nodes and/or are linked to by the same set of nodes.

These new definitions of clusters allow for the formation of groups that are not necessarily densely connected. This is a major break from the traditional notion of a cluster formed by a set of nodes with high density. Partitioning a graph using pattern based clusters often leads to the use of algorithms with high complexity. There are many representative based solutions that allow the reduction of the time complexity of these algorithms, but their performance is closely linked to the selection of representatives. Often, these representatives do not represent a cluster, and nodes that represent small clusters are ignored.

This paper presents a community detection method called Sponge walker. It considers a cluster as a set of nodes with either a high density of interactions or with paths to the same set of nodes. Hence, it considers both a density based definition of a cluster and a pattern based definition. Sponge walker uses a density measure to group nodes together in its first step and merges them together. It then uses a similarity based measure that is more expensive to compute using representatives on the smaller graph obtained in its previous step. These representatives are only used to compute a distance measure between sets of nodes. Hence, the selection of representatives should only be "good enough" as they do not necessarily map to a cluster. This paper is organized as follows: Sect. 2 presents some related works. Sections 3 and 4 introduce the Sponge walker method and show experiments that were conducted on synthetic networks. Section 5 offers some final conclusions.

2   Related Works

This section presents some of the most relevant works. For a complete literature review, refer to [1, 3, 5].

The Informap algorithm [6] considers a cluster as a set of nodes where a random walker could be trapped. It minimizes the minimum description length of random walks performed on the graph.

Pons and Latapy [7] propose a community detection method called Walktrap. They consider a cluster as a set of nodes that can reach or are easily reached by the same set of nodes. Hence, the clusters returned by the algorithm are not necessarily densely connected, although this method can also detect this type of clusters (highly connected nodes can reach and are reached by the same set of nodes). Their method computes at iteration random walk ($t = 4$ by default) on the network. It then computes a distance measure between each pair of nodes using the random walk matrix of step t.

Subelj et al. [8] performed a quantitative and qualitative analysis of several clustering algorithms on real citation networks. They clustered the results of different algorithms using K-Means (11 classes). Clustering results from Walktrap were not put in the same class as clustering results from Louvain algorithm despite optimizing modularity. Hence, despite using modularity to find the best cut, the Walktrap algorithm finds patterns that are not necessarily observed by the Louvain algorithm.

Zhou et al. [9] propose a method for semi-supervised classification that can be used in an unsupervised mode as well. This is achieved by optimizing the smoothening function. This function allows graph patterns that group nodes into communities to be captured by computing similarities using incoming links from hub nodes and outgoing links to authority nodes. Similarly to Pons and Latapy [7], the clusters they find have high similarity with regard to the outgoing and incoming links.

Wang et al. [10] propose a method that distinguishes between nodes with high degrees that represent communities and nodes with lower degrees. They first start by identifying nodes that represent communities using an algorithm called WeBA, and then assign the nodes with lower degrees to fixed communities.

Other methods focus on adapting the modularity measure [4] for directed networks. Arenas et al. [11] propose a modularity measure where the null model takes into account the indegree and outdegree of nodes. Their proposed method nevertheless introduces a bias with regard to the direction of nodes. Flipping the direction of an edge has a high effect with regard to the community which a node is assigned. Kim et al. [12] propose another adaptation for modularity in directed graphs that makes use of random walks. Equation (1) defines modularity in directed graphs where π_i is the stationary distribution for node i. This definition of modularity allows both density based clusters and flow based clusters to be found.

$$\sum_{i,j} [L_{ij} - \pi_i \pi_j] \delta(c_i, c_j) \tag{1}$$

Agreste et al. [13] compared a set of community detection algorithms for directed networks using both synthetic datasets, generated using the generator described in [14], and real datasets. In their experiments, the Walktrap algorithm [7] achieved the highest NMI scores using the synthetic datasets. The lowest scores were achieved when an algorithm for undirected graphs was applied on a transformed directed graph (using Infomap [6] and label propagation [15]). Such transformations aim to capture local patterns stemming from edge directionality (ex co-citation).

Satuluri et al. [16] describe various methods to symmetrize, a graph transforming it from directed to undirected. An algorithm for community detection in undirected graphs could then be used on the obtained graph. One such measure is the use of distances $U = A \cdot A^T + A^T \cdot A$ where $u_{i,j}$ is the number of co-citations and citations from common nodes. One should note that this proximity measure connects nodes that are not connected in the original directed graph. The authors also proposed a degree discounted similarity that increases the proximity between nodes i and j if they have links to nodes with a small indegree or have links coming from nodes with a small outdegree. They use the square root of the degrees rather than the degrees themselves to avoid penalizing too severely co-citations to or from high degree nodes.

Klymko et al. [17] proposed a weighting method for directed graphs to transform them into undirected graphs where any clustering algorithm for undirected graphs could then be used. The method takes into account the number of directed 3-cycles that allow flows of information. Each cycle is additionally weighted by the number of bi-directional it contains.

3 Methodology

The Sponge Walker algorithm is described in Algorithm 1. It starts with a density based optimization and then identifies suitable representatives of clusters in the network. It uses these representatives to find communities based on flow patterns in the network. As opposed to representative based clustering algorithms, it does not consider that representatives capture all the communities. Representative based clustering algorithms usually cluster the representatives and then assign nodes to these representatives. This makes the algorithms highly dependent on finding the right representatives which is not a trivial task.

Algorithm 1 Sponge Walker

1: *do*:
2:      *for each node n*
3:            \\Modularity based clustering
4:                  $y_n \leftarrow assign\_community(n)$
5: *while changes*:
6: create graph of communities
7: *select community representatives*
8:      repeat l times
9:          $S \leftarrow Sample(nodes, k)$
10:      *for each node n_i in S*
11:            $n_m \leftarrow Sample\_indeg\_neighbor(n)$
12:            $w(n_m) \leftarrow w(n_m) + 1$
13:      $Rep \leftarrow max(W, k)$
14: \\Construct Markov chain at t step
15: $M \leftarrow \frac{A}{\sum_j A_{.j}}$
16: $M^{(t)} \leftarrow \prod_t M$
17: \\Compute distances between nodes based on distance to reps
18: repeat
19:      $Dist_{i,C} \leftarrow \sum_k \in Rep \frac{M_{i,k} - M_{C,k}}{indeg(k)}$
20:      merge nodes with shortest mean distance
21: select best partition (max modularity)

Density Based Optimization (Steps 1–6) We use a variation of the Louvain algorithm [18] to group nodes by optimizing a modularity based measure. Three modularity based measures were considered and tested:

The first is a variation of modularity adapted for directed graphs [12]. It makes use of the stationary distribution of a Markov chain computed from the adjacency matrix. This stationary distribution is used to compute the null model. We propose a second variation based on regular modularity but favoring nodes that have both incoming and outgoing edges (reciprocity). This measure loses some of the information linked to the directionality of edges. Nevertheless, in this first step of the algorithm, the objective is only to group densely connected nodes together using an efficient measure. Hence, computing the number of edges between nodes is a suitable indicator. The measure is shown in Eq. (2)

$$Q = \frac{1}{m} \sum_{i,j} \left[A_{ij} + A_{ji} + \sqrt{A_{ij} \cdot A_{ji}} - 3 \cdot \frac{deg_i \cdot deg_j}{m} \right] \delta(c_i, c_j) \qquad (2)$$

This measure has the advantage of favoring nodes with reciprocal interactions, which would have a higher probability of belonging to the same community. It also penalizes neighbors with few edges as their score shrinks because of the additional parameter in the null model (that accounts for reciprocity). Such neighbors would be considered for merging in steps 17–20.

The third approach is to use regular modularity [4] performed by ignoring the edge directionality. As with the classic Louvain Algorithm [18], this modified Louvain algorithm merges nodes that belong to the same community. The difference with Louvain algorithm lies in the fact that the algorithm used merges nodes only once while the Louvain algorithm alternates between optimizing modularity and merging communities. In the experimentations we conducted, the three modularity based measures provided very similar results. This may be due to the fact that only the first iterations of the Louvain algorithm are executed.

In this first phase, the algorithm starts by assigning each node into a separate community. It then iterates over each nodes and assigns it to the community in its neighborhood that maximizes a modularity based measure. The algorithm stops when no more changes are possible or when a maximum number of iterations were executed.

Selection of Representatives (Steps 7–13) These steps concern the selection of nodes (Representatives) that will be treated as features (or dimensions) when computing distances between nodes.

While the previous steps aim at grouping nodes based on whether they have high density of edges, the next steps use pattern based similarities to cluster nodes. They aim to capture the similarity between nodes using the structure of the graph and not simply the local neighborhood as with modularity based measures.

We select representatives because we consider that nodes having similar probabilities of reaching these representatives after t steps should belong to the same cluster. These representatives should hence be far apart (preferably belong to different communities) and be easy to reach (high indegree relative to their neighbors).

The representatives are identified by iteratively sampling nodes at random and sampling a second time one neighbor in the neighborhood of each selected node with a probability relative to its indegree.

The default number of representatives is set to $0.1 \cdot \sqrt{N}$, where N is the number of nodes. The algorithm is stable with regard to variations of the number of representatives. Selecting a large number would nevertheless increase the time complexity of the algorithm and a small number would degrade the performance.

Construction of Markov Chain (Steps 14–16) These steps start by computing the Markov chain for transitions in the social graph using the adjacency matrix. This computation translates into dividing the adjacency matrix by the weights of the outgoing edges for every node (which allows to obtain the Markov chain). Next, the Markov chain at step t is computed. Hence, edges $m_{i,j}$ represents the probability of reaching j from i at step t. When t is very large, the Markov chain approaches the stationary distribution where $M(t + 1) = M^{(t)}$. In this case, the probability of being at node j at step t is independent from where the walk started, making the chain not suited for computing distance between nodes.

The Milgram experiment [19] suggests that there is a chain of length 6 between every two people. Facebook has confirmed that the actual number between facebook users is less then 6. This suggests that in a social network, every user can reach any other user in less than 6 steps. We hence set the t parameter for random walks to 3 to capture patterns in communities rather than global patterns. Note that nodes in this phase represent communities merged in the modularity optimization phase.

Pattern Based Distance (Steps 17–20) This phase concerns computing the distances between every connected node. The distance used is similar to the one presented in Walktrap [7], with the difference that only representatives are used as features. Using representatives as features makes the algorithm much faster without a great loss in the quality of clusters (as representatives are distant enough to capture similarities between nodes). Equation (3) describes the distance computed between a pair of nodes.

$$Dist_{i,j} \leftarrow \sum_{k \in Rep} \frac{M_{i,k} - M_{j,k}}{indeg(k)} \tag{3}$$

Similarly to Walktrap [7], at each step, the nodes with the shortest distance are merged. When two nodes are merged, they are replaced by a node that represents the merged nodes. After merging two nodes, the entry of the Markov chain representing the new community is obtained by averaging the probabilities of the nodes that compose the community.

Selection of the Best Partition (Steps 21) The previous step provides a dendrogram that could be used to obtain several partitions depending on the needs of the user. There are several methods that allow a hard partition to be obtained from a dendrogram. In this study, we select the partition with the maximum modularity.

3.1 Complexity

In order to evaluate the complexity of Sponge Walker, it is necessary to compute the complexity of each one of its components. The worst case complexity for the modularity based optimization cannot be determined, as there is no bound on the complexity of the Louvain algorithm. Experimentation conducted on Louvain suggests that its average complexity is around $O(n \cdot \log(n))$. Also, it is possible to include a condition where the algorithm sets a maximum number of iterations for the modularity optimization phase.

The complexity of the sampling phase is $O(l \cdot |Rep|)$ where l is a parameter and $|Rep|$ is the number of samples.

The computation of distances in the pattern based merging is performed on the graph returned by the Louvain algorithm. In this phase, the worst case occurs when the modularity could not be optimized. Hence, the initialization of the probability matrix (Markov chain at step t) is $O(MNt)$. The complexity of the construction of the dendrogram is $O(M \cdot h \cdot |Rep|)$. The last phase, which includes clustering the representatives can be ignored with regard to time complexity. Hence, it is possible to consider that the worst time complexity of the method is the same as the time complexity of the construction of the probability vector $O(\text{MNt})$. The average time complexity is much smaller, as the modularity optimization phase shrinks the number of edges and nodes in the graph. We estimate the average complexity to be bound by the Louvain algorithm $O(N \log(N))$.

4 Experiments

This section presents experiments that were performed on synthetic datasets. All the datasets were generated using a network generator developed for testing community detection algorithms proposed by Lancichinetti et al. [14]. The first set of experimentations were performed on the default parameters of the network generator. These parameters are the number of nodes (4), the average indegree (15), the minimum community size (20), and the maximum community size (50).

We vary the mixing parameter μ which is the maximum outdegree/indegree ratio of a node. The higher this parameter, the more difficult it is to identify the communities in the network. We use normalized mutual information (NMI) to evaluate the algorithm and we compare it with Walktrap [7], Informap [6] and with Louvain algorithm [18] that we modified to optimize the modularity defined by Kim et al. [12]. We also compare the algorithm with the regular Louvain algorithm when ignoring edge directionality and with Infomap and Louvain after performing a symmetrization of a graph that accounts for edge directionality [16].

Table 1 presents the mean NMI for 100 graphs obtained from the graph generator [14] by varying the seed. It is possible to see that Sponge Walker obtains satisfying results compared to the benchmark algorithms. As the μ parameter increases, community detection becomes more difficult to perform. We could also see that for the first two datasets with a small μ parameter, Sponge walker obtained a high NMI average. Sponge walker was outperformed on the dataset with ($\mu = 0.5$) by the other algorithms, but it nevertheless obtained a high NMI. It outperformed all the benchmark methods on the most difficult dataset, and outperformed all the benchmark algorithms except Walktrap and Louvain on the second hardest dataset. This might be due to the use of the representatives when computing distances which renders the algorithm more stable to noise.

Table 1 Clustering average NMI on a set of graphs with many nodes with high degree

μ	S. W.	Walktrap	Dir. Louv.	Infomap	Louv.	S. Infomap	S. Louv.
0.1	0.96	1	1	1	1	0.97	1
0.25	0.97	0.99	1	1	1	0.92	0.89
0.50	0.96	0.99	0.99	1	0.99	0.85	0
0.75	0.53	0.72	0.01	0	0.60	0.187	0
1	0.33	0.06	0	0	0.04	0.06	0

Table 2 Clustering average modularity on a set of graphs with many nodes with high degree

μ	S. W.	Walktrap	Dir. Louv.	Infomap	Louv.	S. Infomap	S. Louv.
0.1	0.81	0.90	0.90	0.90	0.89	0.89	0.89
0.25	0.65	0.71	0.71	0.71	0.67	0.60	0.66
0.50	0.41	0.46	0.46	0.46	0.43	0	0.41
0.75	0.10	0.18	0.01	0	0.18	0	0.04
1	0.08	0.1	0	0.15	0	0	0.08

The second set of experiments were performed using the same network generator [14]. The parameters used are the number of nodes (1000), the average indegree (5), the minimum community size (200), and the maximum community size (300).

This second set is conducted on sparser graphs with larger communities as compared with the first set of experiments. Table 2 presents the results obtained. The average NMI scores on this dataset are lower due to the sparse nature of the networks. Louvain on the symmetrized network has the highest performance but also the highest complexity. It performs well compared to density based methods as the networks are very sparse. Sponge walker has the second best performance on graphs with high μ parameter which are more difficult to cluster. Note that Sponge walker obtains satisfying results despite having a much lower complexity than Walktrap.

5 Conclusion

Community detection has become one of the central subjects in social network analysis. It has a variety of applications such as exploratory data analysis, recommendation systems, or anomaly detection.

Community detection in social networks has traditionally represented networks as undirected graphs. Hence, the majority of community detection algorithms make the assumption that an interaction is reciprocal. Pattern based or flow based community detection in directed graphs allows leveraging of the patterns of communication and could hence return richer results. These methods are often much more expensive compared to density based community detection. With the availability of large graphs, pattern based algorithms with a complexity of $O(N^2)$ or more become non-desirable.

This paper presents an algorithm that uses efficient density based community detection in its first phase to capture local patterns. It then leverages the local patterns it found and performs a pattern based community detection using representatives. This allows the algorithm to provide satisfying results and to run in $O(n \cdot \log(n))$ on average.

Acknowledgements This work was supported by the French Investment for the future project REQUEST (REcursive QUEry and Scalable Technologies) and the region of Champagne-Ardenne.

References

1. Malliaros, F.D., Vazirgiannis, M.: Clustering and community detection in directed networks: a survey. Phys. Rep. **533**(4), 95–142 (2013)
2. Erdos, P., Renyi, A.: On the evolution of random graphs. Publ. Math. Inst. Hung. Acad. **32**(4), 1312–1315 (1960)
3. Fortunato, S.: Community detection in graphs, pp. 75–174. http://www.sciencedirect.com/science/article/pii/S0370157309002841 (2010)
4. Newman, M.: Modularity and community structure in networks. Proc. Natl. Acad. Sci. USA **103**(23), 8577–8582 (2006). http://www.pubmedcentral.nih.gov/articlerender.fcgi?artid=1482622&tool=pmcentrez&rendertype=abstract
5. Schaeffer, S.E.: Graph clustering. Comput. Sci. Rev. **1**(1), 27–64 (2007)
6. Rosvall, M., Bergstrom, C.T.: Maps of random walks on complex networks reveal community structure. Proc. Natl. Acad. Sci. USA **105**(4), 1118–1123 (2008). http://www.ncbi.nlm.nih.gov/pubmed/18216267
7. Pons, P., Latapy, M.: Computing communities in large networks using random walks. Lect. Notes Comput. Sci. **3733**, 284–293 (2005)
8. Subelj, L., Van Eck, N.J., Waltman, L.: Clustering scientific publications based on citation relations: a systematic comparison of different methods. PLoS One **11**(4), 1–23 (2016)
9. Zhou, D., Schölkopf, B., Hofmann, T.: Semi-supervised learning on directed graphs. Adv. Neural Inf. Proces. Syst. **17**, 1633–1640 (2005)
10. Wang, L., Lou, T., Tang, J., Hopcroft, J.E.: Detecting community kernels in large social networks. In: Data Mining (ICDM), pp. 784–793 (2011)
11. Arenas, A., Duch, J., Fern, A.: Size reduction of complex networks preserving modularity. New J. Phys. **9**(6), 176 (2007)
12. Kim, Y., Son, S.W., Jeong, H.: Finding communities in directed networks. Phys. Rev. E Stat. Nonlinear Soft Matter Phys. **81**(1), 1–9 (2010)
13. Agreste, S., De Meo, P., Fiumara, G., Piccione, G., Piccolo, S., Rosaci, D., Sarne, G.M.L., Vasilakos, A.: An empirical comparison of algorithms to find communities in directed graphs and their application in Web Data Analytics. IEEE Trans. Big Data **3**(3), 289–306 (2016). http://ieeexplore.ieee.org/document/7755743/
14. Lancichinetti, A., Fortunato, S.: Benchmarks for testing community detection algorithms on directed and weighted graphs with overlapping communities. Phys. Rev. E Stat. Nonlinear Soft Matter Phys. **80**(1), 1–9 (2009)
15. Raghavan, U.N., Albert, R., Kumara, S.: Near linear time algorithm to detect community structures in large-scale networks. Phys. Rev. E **76**, 036106 (2007). http://arxiv.org/abs/0709.2938
16. Satuluri, V., Parthasarathy, S.: Symmetrizations for clustering directed graphs. In: Proceedings of the 14th International Conference on Extending Database Technology, no. i, pp. 343–354 (2011). http://doi.acm.org/10.1145/1951365.1951407
17. Klymko, C., Gleich, D., Kolda, T.G.: Using triangles to improve community detection in directed networks (2014). http://arxiv.org/abs/1404.5874
18. Blondel, V.D., Guillaume, J.-L., Lambiotte, R., Lefebvre, E.: Fast unfolding of communities in large networks. J. Stat. Mech. Theory Exp. **10008**(10), 6 (2008). http://arxiv.org/abs/0803.0476
19. Stanley, M.: The small world problem. Pshychol. Today **1**(1), 61–67 (1967)

Part IV
Algorithms and Applications II

Market Basket Analysis Using Minimum Spanning Trees

Mauricio A. Valle, Gonzalo A. Ruz, and Rodrigo Morrás

Abstract Marketing efforts and store layout could benefit from studying purchases that commonly happen together. This type of studies are commonly referred to as market basket analysis (MBA). In this work, a market basket methodology based on minimum spanning trees (MST) is presented. Because of the wide variety of products in a typical grocery store, and the heterogeneity of consumer shopping behavior, MBA is a complex task, from a computational point of view, and for subsequent interpretations of the results. The proposed methodology simplifies significantly the process of finding sets of products that have high co-occurrence in the market basket of the consumers, that is, products that are bought together. The resulting MST as a visual representation that connects all products with a high correlation to each other is easy to interpret and becomes a powerful tool to propose marketing actions. This solution turns out to be a complement with the traditional association rules used for MBA.

Keywords Market basket analysis · Minimum spanning tree · Network of products · Association rules

M. A. Valle (✉)
Facultad de Economía y Negocios, Universidad Finis Terrae, Santiago, Chile
e-mail: mvalle@uft.cl

G. A. Ruz
Facultad de Ingeniería y Ciencias, Universidad Adolfo Ibáñez, Santiago, Chile
e-mail: rmorras@uai.cl

R. Morrás
Escuela de Negocios, Universidad Adolfo Ibáñez, Santiago, Chile
e-mail: gonzalo.ruz@uai.cl

© Springer International Publishing AG, part of Springer Nature 2018
R. Alhajj et al. (eds.), *Network Intelligence Meets User Centered Social Media Networks*, Lecture Notes in Social Networks,
https://doi.org/10.1007/978-3-319-90312-5_11

1 Introduction

Determining promotion strategies and layout to maximize sales is a permanent challenge for retail chain managers. For example, what products should be promoted together? what is the greatest product supply on a shelf to increase the chances of them being part of a market basket? what products are complementary within a product range of the same category? The answers to such questions may be found through market basket analysis (MBA). The MBA is a data mining method focused on finding consumer buying patterns in transactional databases [14]. A transactional database contains at least consumer identities or identifiers, the product (or item) set purchased (which constitutes the market basket), and the amount of each item purchased. Thus, the MBA seeks to find product associations, i.e., identify which products are usually bought together and which ones are not.

The data mining tool for which the MBA has several applications in retail marketing has traditionally been association rules mining. Association rules (ARs) [1] make it possible to find such statements as: *when a person buys spaghetti, they also buy tomato sauce*. This means, an antecedent is identified: *the spaghetti purchase*, or the left side of the rule, and a consequent: *the tomato sauce purchase*, or the right side of the rule. The popularity of association rules lies in the explicit way in which the association between products (antecedent and consequent) is established, which is why it is easy to interpret in order to take actions that allow, for example, offers of tied products to be changed or modified or targeted promotions to be changed to increase customer loyalty.

The greatest disadvantage of the association rules is that in practice the algorithm usually finds a number of rules so numerous that analyzing those of interest to the analyst or manager is usually intensive and time-consuming [11]. Even though there are several indices available to assess the quality of the rules found, when they are ordered by different measures, ranking ordered rules according to their quality varies, and it is therefore currently not possible to have an objective measurement to determine which rules are better than others in their ability to explain purchasing behavior.

This study proposes an complementary alternative to the ARs, which allows a visual and fast way to find product associations in terms of a *product network* [18, 19] based on minimum spanning trees (MST). In other words, the network topology by itself makes it possible to identify associations between product pairs, and consequently to reduce the search space of association rules to only those that the network reveals.

This study presents a MST-based methodology that can identify significant product associations in a sample product basket, intuitively and clearly, discarding potential spurious rules and leaving only those with a high level of dependency.

2 Concepts and Definition of the Problem

2.1 Association Rules

Basket data represents the set of n items or products available to the consumer. It is defined as $I = i_1, \ldots, i_n$. The association rules try to find dependency between items that comprise consumer market baskets and which are found in a transactional database D. Support is defined as the degree of popularity of an item, and it is measured as the proportion of transaction in which the item appears. In other words, it represents the probability that the item i_k is present, i.e., $Support(i_i) = P(i_i)$. Confidence expresses the likelihood that an item i will be taken (called consequent) when item k is also purchased (antecedent), and is measured as the proportion of transactions with item i_i in which item i_k also appears, i.e., $Confidence(i_i \rightarrow i_k) = P(i_i | i_k)$. Finally, Lift measures how likely it is that item i_i will be purchased when i_k is also purchased, controlling for the popularity of item i_k. $Lift(i_i \rightarrow i_k) = P(i_i | i_k)/P(i_k)$. If the lift of a rule is greater than 1, then the occurrence of the antecedent and the consequent are dependent upon one another, and therefore the rule is potentially useful in predicting the consequent in future transactions. Specifically, it is said that $i_1 \rightarrow i_2$, when:

1. i_1 and i_2 occur together at least $s\%$ of the n market baskets. The rules found must have at least this level of support,
2. All the baskets that contain i_1, at least $c\%$ also contain i_2. The rules found must have at least this confidence level.

The parameters $s\%$ and $c\%$ are given by the analyst. Under these two conditions, all the rules found that fulfill them will be subjected to analysis. As previously indicated, the main limitation of ARs is that for a transactional database, the number of rules generated fulfilling the two previous conditions may be hundreds or thousands, which, although being ordered by lift or another quality measurement, it is difficult to analyze them all or find rules that are particularly interesting for retail managers.

2.2 Product Networks

Product networks can be constructed from a list of transactions. Each node represents an item and the edges that connect each vertex of the network represent whether the pair of items were purchased in the same transaction [19]. This is the intuitive description of the product network; however, similarly to what occurs in many other phenomena represented by networks, it is observed that there is a small set of nodes or vertices that have a high connectivity (to other nodes), i.e., a high degree, whereas many nodes have a very low connectivity (low degree). This means that the distribution of the degree of the network is heavily skewed to the right, resulting in the typical scale-free networks [3].

The high skewness in the distribution of the product network imposes practical problems because for product networks, the existence of an edge does not necessarily mean that two products are present in each market basket. The presence of some products in a basket could not be explained due to an association with another product, but simply because it is a circumstantial purchase because of an external factor outside of the habitual buying pattern. For example, a household buyer with a regular shopping list in his mind could include in his basket the product X on behalf of a third person. This product is not associated with the basket of habitual products of this buyer. However in the network, the item X is now connected with the rest of the products in the basket.

Additionally, it is easy to imagine that dealing with hundreds of products in a supermarket, the potential number of edges is enormous. In a network with just n nodes, there are potentially $0.5n(n-1)$ edges. For example, with only 200 products, there are potential 19,900 edges, which makes it very difficult to find buying patterns and as a result it is impractical. To overcome this problem, several alternatives have been considered. For example, product community detection is widely known in networks and has been used in various fields of application, particularly in genetics [5, 8, 16]. In product networks, *information density* has been used as a parameter to find product communities [19], which measures the degree of information present in a community. Thus, the more information a product group has, the more susceptible this group is to becoming a community.

Another apparently effective and simpler alternative is the application of thresholds to the network weight edges [20], i.e., those edges with weights below a determined value are removed from the network, leaving only those edges that present with heavier weights. Thus, it is possible to find and describe product zones with a strong relation between them.

Although these market basket analytical methods using product networks manage to capture valuable information regarding the relationships that exist between a wide range of items, they still lack a visual and practical representation for use in the field and for making decisions in situ. Additionally, the commonly used approaches are variations of other solutions applied in other disciplines, whereas this study offers an ad-hoc solution to the problem of market baskets, considering that, in essence, this problem is one of correlation identifications [7].

2.3 Minimum Spanning Trees

The minimum spanning trees (MSTs) are a special type of graph in which all the nodes are connected without forming cycles. This way, the n vertices in the network will have at the most $n-1$ edges. Unlike a product network in which the edge represents the presence or absence of products in the same market basket, in the MST, the edges represent the degree of correlation between items, which is then suitably transformed to a distance measurement.

MSTs have been used to study the interactions existing in the currency market. For example, they have been used to study the systemic behavior of share prices on the US stock market [6]. The foreign currency market has been studied as an MST network, finding groupings of currencies highly related to each other. Recently, methods to filter information have been used to model correlations between share prices on the German DAX, creating visualizations of correlation-based networks [4, 17]. These are only some examples of the potential that MSTs have to gain greater understanding of market behavior. In this work, the phenomenon is consumer purchase decisions. Evidently, when dealing with a large number of nodes (products), one might think about the need to apply some type of filter with the aim of removing links with low correlation (long distance between nodes). However, one advantage of the MST is that such networks show only the most important nodes (products), creating a compact network. Moreover, the MST allows us to find the hierarchical organization of the nodes [15] in such a way that it is possible to discover homogeneous groups of categories or products with respect to consumer preferences in the total product set offered by the supermarket.

The network is based on correlations (and not merely on the presence of products in the same basket), that is, there are positive and negative associations between pairs of items, which are translated into a metric distance. As a result, an MST is built from which allows us to determinate the set of products with high co-occurrence with each other. This way, a topology is obtained that represents a summarized form of a large number of transactions at a level that can be managed or interpreted by a manager.

3   Methodology

The proposed market basket analysis methodology is summarized in the following three steps:

1. Obtain a correlation matrix between products.
2. Transform the correlations into a measurement of distance.
3. Obtain the MST.

The first step consists of obtaining a measurement of association between basket products. The transactional database keeps a record of the products taken in each transaction, so that the presence or absence of the product in a basket is represented as a binary variable. Thus, the ϕ correlation index is calculated [12], which represents the degree of association between binary variables. For n products, the result is a symmetric $n \times n$ matrix. Just as in the Pearson correlation, the ϕ correlation takes values between -1 and 1. More simple measures could be used, for example simply by counting the number of basket in which the two products co-appear. However, this kind of measures does not indicate association between products. Then it is necessary to transform these correlations into a metric of distance, so that when the correlation is equal to 1, the distance is zero. One transformation that fulfills this criterion and axioms of metric distance [15] is:

$$d_{ij} = \sqrt{(2(1 - \phi_{ij}))} \tag{1}$$

where ϕ_{ij} is the ϕ coefficient between items i and j. The correlation matrix is now transformed according to the expression (1). The symmetry is maintained, i.e., $d_{ij} = d_{ji}$, with null values on the diagonal. The third step consists of identifying the MST from the distance matrix. This is carried out with Prim's or Kruskal's algorithm [9]. Finally, a hierarchical grouping process is applied, from which it is possible to detect agglomerations or groups of products. The foundation of this grouping is based on a product j in the MST being at a distance from another product i in such a way that the entire set of products that are at a very short distance from each other, but distant from another group, are candidates to form a product cluster. The practical significance of the clusters found is relevant because they reveal the set of items that contains complementary products, i.e., that are usually purchased in the same transaction.

4 Results

4.1 Data Exploration

As an example application, a real transactional database is taken, containing 1,046,804 transactions and 220 subcategory items from a typical supermarket in a 15-month period. To get an idea of the variability and complexity of the transactions, a sample of 80 subcategory items has been extracted to represent a heatmap of the transactions (see Fig. 1). The red color represents the presence of the item in the transaction, whereas the grey color represents its absence.

Figure 1 illustrates the complexity of the market basket analysis due to the variety of products and particularly to the heterogeneity of what is in each basket. The proposed methodology helps reduce the complexity of the data so as to describe significant buying patterns. A simple frequency analysis of items in the market basket reveals that the products most frequently bought are item 80 (vitaminized pasta), 214 (beaten yogurt), and 7 (vegetable oil).

Figure 2 clearly shows those products that have the greatest turnaround in the supermarket, which may be of use in establishing strategies for products with the highest sales volumes. However, this basic information does not say anything about the relationship that these products have with others present in the same basket. For example, it could be that a product not purchased very frequently is taken with almost total certainty with some other complementary product. From the promotional strategy point of view, the profitability of these products can be much greater than that of another product that is more frequently purchased.

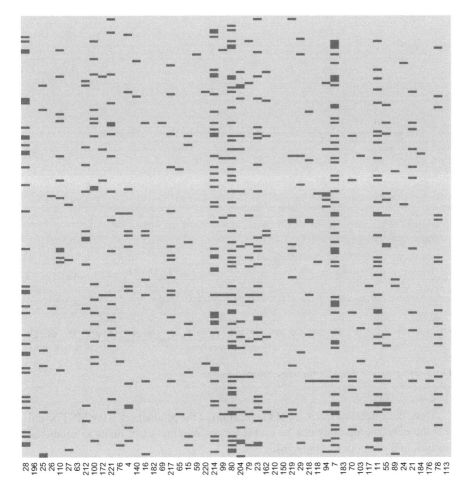

Fig. 1 Heatmap of a random sample of 200 transactions and 80 subcategory items

MST Result

After finding correlations between all pairs of subcategory items, and convert them to distances according to Eq. (1), an MST emerges (see Fig. 3). The MST found reflects at least 15 product branches that emanate from the main stem, which means potential item cluster candidates. On the other hand, the edges that connect each node being shorter indicates that items have a high likelihood to be purchased at the same time. It is interesting to observe that the items of the same subcategory tend to group together in the same branch. For example, wines products (pink nodes) are located to the right branch of the MST, while almost all the cleaning house articles (magenta) are located in the lower left part of the MST. Other products like milk (blue nodes), yogurt (green nodes), and cheese (cyan nodes) are located together in the upper part of the MST, over four branches.

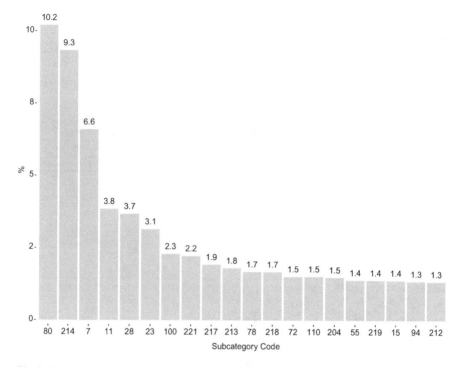

Fig. 2 Frequency of items for the transactional database

Thus, the careful observation of the location of the product set in the MST reveals that the network topology can be useful for discovering set of items as candidates for bids and promotion campaigns.

MST as a Good Associations Rules Discoverer

The MST has 220 nodes or category products, and therefore 219 edges. Each of these edges can be considered as an association rule. As the MST includes low-distance connections, i.e., high level of correlation between products, these connections are equivalent to association rules with high levels of Lift.

To demonstrate that the MST easily delivers high-quality rules, we developed a simulation in which for each MST product i, we search for the set R_i of all association rules of type $P_i \rightarrow P_j$ where $i \neq j$. That is, for our case, there will be 220 different sets of rules. Then, for set R_i we find the rule $P_i \rightarrow P_m$, where m represents the product or node that is connected to the product i in the MST. For this rule we obtain their respective Lift. Then, we compare this Lift level with the Lift mean of the set rules R_i. This procedure is carried out for all products.

The result of the analysis is shown in Fig. 4. We plotted the Lift of the association rules given by the MST. There is also the percentile 75 and 90 of the distribution of

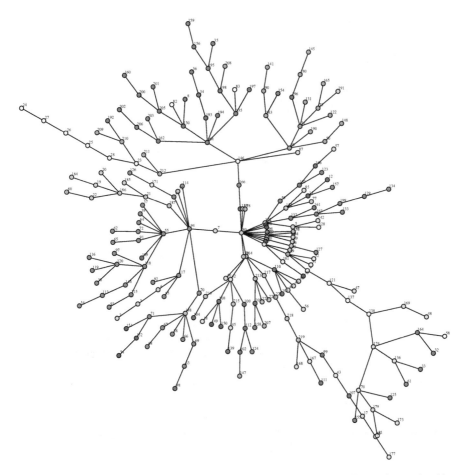

Fig. 3 Representation of the complete MST for the transactional database. The numbers at the side of each node is the subcategory to which the item belongs to. Each color represents a subcategory: white—red meat; red—bird meat; yellow—white meat; blue—milk; green—yogurt products; cyan—cheese; brown—rice; light green—oats and cereals; yellow/green—noodles; gray—oil; khaki—soft drinks; pink—wines; magenta—cleaning articles; light blue—personal hygiene items; dark blue—baby items; deep pink—butter and delicacies; orange—others

all Lifts for each set of rules R_i. It is easy to realize that the association rules from the MST are of high quality. All these rules, except 4, have a Lift level above the 75th percentile, which means that these rules are almost always within the 25% of rules with higher level of Lift. It is even possible to observe that most of the rules of the MST are above the 90th percentile. In fact only 40 rules of 219 are below the orange line (18%). These results give us assurance that the MST is also a graphical representation of association rules with high quality.

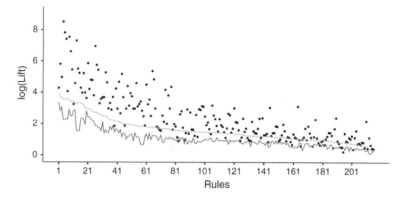

Fig. 4 Plot of log(Lifts) of the association rules given the MST (black points). The red and orange lines represent the percentile 75 and 90, respectively, of the distribution of log(lift) for each set R_i of association rules

Example of Application

In order to demonstrate the simplicity of the proposed methodology in contrast to association rules, a process of rules discovery was carried out using the Arules algorithm [2, 10]. Using a minimum support and minimum confidence of 0.001 and 0.01, respectively, 179,610 association rules were found. To get an idea of the complexity of the analysis of association rules, Fig. 5 shows a scatter plot of the rules with lift greater than 1, which totals 33,759, i.e., although the number of rules has been significantly reduced, the number of rules remains too high to perform a practical analysis and recommend promotion or marketing strategies.

To observe the set of association rules in greater detail, Table 1 presents eight rules (of the 33,759) with the highest level of lift. For example, the antecedent of the first rule is subcategory 22 (balsam conditioner), and the consequent is 186 (specific shampoo). The interpretation of this rule is that when balsam conditioner is taken, then specific shampoo is also purchased. This rule is easily deduced from the MST (Fig. 3, lower part of the MST, light blue nodes). In fact, for example, the rule 51 → 189 (51 represents preserved cholga which is a kind of seafood, and 189 represents seafood broth) is also easily detected in the MST, to the left side from the center of the MST.

For rules with more than two products, the location of the nodes in the graph is harder to find and does not appear to be trivial relationships. For example, rule number 4 involves personal care products (118, 94) and cleaning products (187, 21). Each pair of products is close to each other, but more distant among them in the MST because they are located in different branches. This is not surprising since the construction of the MST is based on correlations of product pairs, and does not take into account conditional correlations.

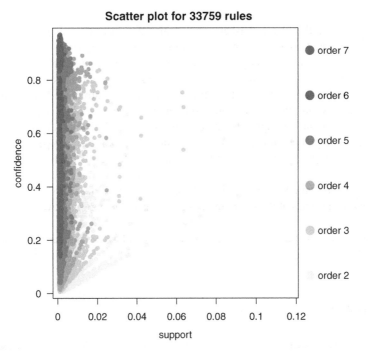

Fig. 5 Scatter plot of association rules with lift greater than 1. On the right side of the plot the level of lift is shown according to the intensity of the color. The vertical axis indicates the confidence level of the rules

Table 1 First 8 rules found with the highest level of lift

	Left-hand side (LHS)	→	Right-hand side (RHS)	Support	Confidence	Lift
1	22	→	186	0.0025	0.314	20.018
2	51	→	189	0.0011	0.215	13.627
3	80, 118, 94, 187	→	21	0.0010	0.565	13.467
4	118, 94, 187	→	21	0.0013	0.552	13.156
5	214, 11, 21, 72	→	187	0.0010	0.756	13.148
6	80, 94, 21, 72	→	187	0.0011	0.755	13.138
7	214, 80, 7, 21, 72	→	187	0.0011	0.752	13.091
8	94, 7, 21, 72	→	187	0.0011	0.750	13.042

5 Conclusions

The MST is a graphic representation of the role that the main items in a supermarket basket play, and thus also the items that have a peripheral role in influencing the market basket. The nodes that possess a degree greater than two are considered to be more important due to their greater influence on the market basket. As observed in the application, it was possible to quickly find those products with a high propensity for inclusion in the market basket in combination. Therefore, this tool promises to be an excellent alternative to propose in situ marketing strategies.

The next stage in this study is to build MST based on conditional correlations (or partial correlations). Sometimes some correlations between pairs of products can be given due to the influence of a third product. Taking into consideration this, the inferred edges can represent a more direct influence between the vertices [13].

Also, we are able to apply the methodology to a transactional database of Latin American companies. In particular, information is available on a supermarket chain in the different geographic areas of Santiago, Chile. It is expected that the MST will be able to identify different purchasing behaviors among the different branches of the same supermarket chain, which will then make it possible to establish offers and promotion strategies at the *local* and not necessarily the national level.

Acknowledgements This work was supported by Fondecyt Research Scholarship (Chile), grant No: 11160072.

References

1. Agrawal, R., Srikant, R.: Fast algorithm for mining association rules in large database. Tech. rep., Research report RJ 9839, IBM Almaden Research Center, Santiago, Chile (1994)
2. Agrawal, R., Srikant, R.: Fast algorithm for mining association rules in large database. Customer Contact World Magazine, pp. 26–31 (2000)
3. Barabasi, A.: Scale-free networks: a decade and beyond. Science **325**(5939), 412–413 (2009)
4. Birch, J., Pantelous, A.A., Soramaki, K.: Analysis of correlation based networks representing dax 30 stock price returns. Pers. Psychol. **47**(4), 501–525 (2016)
5. Blondel, V.D., Guillaume, J.L., Lambiotte, R., Lefebvre, E.: Fast unfolding of communities in large networks. J. Stat. Mech: Theory Exp. **2008**(10), P10008 (2008). http://stacks.iop.org/1742-5468/2008/i=10/a=P10008
6. Bonanno, G., Caldarelli, G., Lillo, F., Mantegna, R. N.: Topology of correlation-based minimal spanning trees in real and model markets. Physical Review E. **68**(4), 046130 (2003). DOI:10.1103/PhysRevE.68.046130
7. Brin, S., Motwani, R., Silverstein, C.: Beyond market baskets: generalizing association rules to correlations. In: Proceedings of the 1997 ACM SIGMOD International Conference on Management of Data, vol. 26, pp. 265–276. ACM, New York (1997)
8. Clauset, A., Newman, M.E.J., Moore, C.: Finding community structure in very large networks. Phys. Rev. E **70**(6), 066111 (2004)
9. Graham, R.L., Hell, P.: On the history of the minimum spanning tree problem. Ann. Hist. Comput. **7**(1), 43–57 (1985)
10. Hahsler, M., Hornik, K., Reutterer, T.: Implications of probabilistic data modeling for mining association rules. In: From Data and Information Analysis to Knowledge Engineering, pp. 598–605. Springer, Berlin (2006)
11. Klementtinen, M., Mannila, H., Ronkainen, P., Toivonen, H., Verkamo, A.I.: Finding interesting rules from large sets of discovered association rules. In: Proceedings of the Third International Conference on Information and Knowledge Management, pp. 401–407 (1994)
12. Knobbe, A.J., Adriaans, P.W.: Analysing binary associations. In: Proceedings of KDD, vol. 96, p. 311 (1996)

13. Kolaczyk, E.D., Csárdi, G.: Statistical Analysis of Network Data with R. Springer, Berlin (2014)
14. Linoff, G., Berry, M.: Data Mining Techniques: For Marketing, Sales, and Customer Relationship Management. Wiley, New York (2011)
15. Mantegna, R.N.: Hierarchical structure in financial markets. Eur. Phys. J. B **11**(1), 193–197 (1999)
16. Newman, M.: Finding community structure in networks using the eigenvectors of matrices. Phys. Rev. E **74**(3), 036104 (2004)
17. Onnela, J.P., Kaski, K., Kertész, J.: Clustering and information in correlation based financial networks. Eur. Phys. J. B **38**(2), 353–362 (2004)
18. Raeder, T., Chawla, N.V.: Modeling a store's product space as a social network. In: International Conference on Advances in Social Network Analysis and Mining, pp. 164–169. IEEE, New York (2009)
19. Raeder, T., Chawla, N.V.: Market basket analysis with networks. Soc. Netw. Anal. Min. **1**(2), 97–113 (2011)
20. Videla-Cavieres, I.F., Ríos, S.A.: Extending market basket analysis with graph mining techniques: a real case. Expert Syst. Appl. **41**(4), 1928–1936 (2014)

Simulating Trade in Economic Networks with TrEcSim

Gabriel Barina, Calin Sicoe, Mihai Udrescu, and Mircea Vladutiu

Abstract Motivated by the large-scale applicability of complex networks, we propose a novel socioeconomic simulator inspired by empirical observations and state-of-the-art economic models. As such, our Trade and Economic Simulator (TrEcSim) is able to use any fundamental complex network topology as an underlying exchange network, and it also introduces a novel heuristic approach to drive the behavior of economic agents, according to theories pertaining to main schools of economic thought. Our simulation results indicate that TrEcSim is a valuable tool for simulating the dynamics of trade in economic networks. Indeed, our simulation results indicate a correlation between the topological properties of the economic exchange networks and the distribution of total payoff: for random and small-world the distribution is meritocratic, whereas for scale-free networks it is topocratic.

Keywords TrEcSim · Simulator · Heuristic · Economic agents · Complex networks

1 Introduction

Complex networks cover an ever-increasing area of scientific research, as they are mainly underpinned by empirical studies of many real-world systems, such as—

G. Barina (✉)
Department of Computers and Information Technology, "Politehnica" University of Timişoara, Timişoara, Romania

Faculty of Automation and Computers, University Politehnica of Timisoara, Timisoara, Romania
e-mail: gabriel.barina@cs.upt.ro

C. Sicoe · M. Udrescu · M. Vladutiu
Department of Computers and Information Technology, "Politehnica" University of Timişoara, Timişoara, Romania
e-mail: mudrescu@cs.upt.ro; mircea.vladutiu@cs.upt.ro

© Springer International Publishing AG, part of Springer Nature 2018   169
R. Alhajj et al. (eds.), *Network Intelligence Meets User Centered Social Media Networks*, Lecture Notes in Social Networks,
https://doi.org/10.1007/978-3-319-90312-5_12

but not limited to—social or economic networks [23]. To this end, social networks can be modeled as complex graph structures made up of a set of social actors or agents (individuals or well-defined groups), linked together via a complex set of ties defined according to the interactions between agents. Therefore, the social agents' connections represent the convergence of its multitude of social contacts [13, 22]. Social networks infuse our social and economic lives profoundly, playing an important role not only in our personal life, but also in our professions, by acknowledging and transmitting critical information about job or trade opportunities [16]. As such, the economic agents, represented as nodes or vertices in the economic networks, can be analyzed according to their topological position in the network, for example, some of them can be very well connected while others may be completely isolated, and so forth.

Social networks are extensively studied in most traditional social sciences but have received an increased boost of interest only over the last couple decades, due to the tremendous development in big data techniques and technologies, including network science [15]. Indeed, network science brings a better understanding for the structure and behavior of social and economic networks, thus proving that human interaction is not only important in social science, but it is also essential for many other fields such as technology and engineering.

In the field of economics, one important issue is to understand why some economic agents fare better than others, or what type of social and economic networks might be better suited for the benefit of the entire society in comparison to the ones we currently have [16]. However, these aspects, along with many others, can only be determined by either past outcomes or qualitative estimations, and not by real-time quantitative observations, mainly due to the fact that economic networks are nonlinear, unpredictable complex systems [3, 6]. To this end, we can use mathematical models or economic simulators that are meant to predict certain behavior of existing economic agents, or even global outcomes, based on algorithms which themselves are based on a multitude of past observations, or on existing economic theories and models. Most previous approaches present specific drawbacks, either due to simulation restrictions (such as the limitation of the simulation algorithm to a static model), the absence of an adequately complex algorithm to drive the behavior of agents, or due to certain hardware requirements.

Therefore, TrEcSim[1] is set to alleviate some of the limitations which characterize previous approaches, as pointed above. To this end, TrEcSim takes into consideration a broad range of customizable parameters, scenarios, or even economic theories. This allows its users to either import certain network topologies—that can be created and analyzed with any of the existing tools, for example, Gephi [5, 17]—or create a basic one using the simulator's interface, and subsequently use it as a starting point in order to simulate diverse scenarios with its basic (i.e., default) settings. Therefore, we use our simulator to explore the connection between economic network topology models and payoff distribution.

---

[1]TrEcSim is freely available at https://github.com/trecsim/trecsim.

In order to better explain the rationale behind our statements, objectives, and results, we organized the rest of the paper as follows: Sect. 2 describes the state of the art in the domain of economic simulation algorithms, Sect. 3 presents the driving idea behind our approach, followed by experimental results, as presented in Sect. 4, leaving the last section of the paper for the conclusions.

2 Background

Visual representation for analyzing the dynamics of social and economic networks fosters data analysis and, at the same time, facilitates a convenient representation of analysis results. As such, many of the existing simulation tools offer the possibility of visualizing the simulated economic networks, as well as the obtained results. Data exploration is performed through displaying nodes and links using a variety of layouts and by attributing colors and size to nodes/agents, according to certain relevant network properties and centralities (e.g., modularity, degree, betweenness, etc.). Visual representations of networks may be a powerful method for conveying complex information, but care should be taken in interpreting node and graph properties from visual displays alone, as they may misrepresent structural properties; indeed, such structural network properties are better captured through statistical, quantitative tools. Nonetheless, the typical application for any economic simulator is to visualize, analyze, evaluate, or verify specific economic scenarios, or theoretical economic models.

One of the most interesting mathematical approaches that has recently received attention from the research community is the *rockstar model* described by Borondo et al. [7]; in this paper, the authors explore how the revenue is distributed in an economic network, according to the density of network ties (i.e., network's links or edges). To this end, the authors of [7] have separated the income/revenue of economic agents into two distinct sources, namely the revenue the agents acquire from their own production, and the revenue received via their role of trade intermediaries. However, their approach is mainly a static model, because the agent's role as producer/rockstar or intermediary is fixed, while the value of their products is also fixed. Moreover, their theoretical analysis considers only on a single transaction cycle for each agent in the network, therefore cannot be labeled as truly dynamical. Another drawback is that, in order to make use of statistical tools, reference [7] assumes only the random topology for the underlying network of economic agents; however, empirical examples show that economic trade networks are not random, as they exhibit small-world and scale-free network properties [9, 24].

As our present paper's approach tries to address the limitations presented above, the first issue consists of considering the value of a given product as not being fixed. One of the most notable theories of value in economic science is the theory of marginality, which examines the increase in satisfaction consumers gain from purchasing and consuming an extra unit of a given good. This theory also attempts to explain the value differences for goods and services by reference to their secondary, or marginal, utility [19, 21]. Conversely, the labor theory of values, usually associated with Marxian economics, describes the effective (economic) value of a given commodity as being determined by the total amount of labor (measured in time and effort) required to produce it, rather than by the use or pleasure its owner receives from it [14, 20].

Currently, to the best of our knowledge, there are no simulators to fully support any of the economic theories or mathematical models that are described above, but there are a few niche simulation tools that are employed for visual representation of dynamic economic networks. However, besides the mentioned flaws, the available simulators have other limiting factors, such as limited simulation options, lack of complex algorithms for modeling economic behavior, or the necessity of supercomputers. Nonetheless, a handful of such simulation software tools worth mentioning in the context of our paper are

– Minsky: the open-source visual computer software for building and simulating dynamic and economic models, which is mainly used in accounting [18]
– Ecolego: a computer simulator used for creating dynamic models and performing probabilistic simulation [4, 8]. By interacting with its GUI (graphical user interface), users can define a handful (but limited) parameters and simulation settings. Ecolego also helps to create reports and to plot simulation results
– EMINERS: a computer simulation software for quantitative mineral resource assessment written in C++ [12]. Though initially it was capable of analyzing data for costs of labor and raw materials, costs for improving mining techniques and their (economic) advantage, as well as several beneficiation methods, it only took a handful of years until EMINERS' algorithms and modules became outdated.
– EURACE: an agent-based model capable of simulating not only single industries, markets, or communities, but also the economic activities at the European Union level. The presented model is designed to factor in artificial markets for real commodities (e.g., consumption goods, investment goods, and labor) when simulating a new economic scenario, as well as financial assets (e.g., debt securities, bonds, and stocks). EURACE yields results that are identifiable in our day-to-day economic activities, by running large-model simulations, which require massively parallel computing on large supercomputers that are not available to the general public [10, 11]

3 Model Description

The main objective for TrEcSim is to provide a simulation framework that is capable of supporting an improved, dynamic counterpart for Borondo's static mathematical model [7]. As such, TrEcSim is developed using the ASP.NET framework, which offers multiple advantages right out-of-the-box. Our application also has the ability of importing network-related data (topology, node-, and edge-count, link-configuration, etc.) from an external .csv (comma-separated values) file, created by one of the available third-party applications (e.g., Gephi [5, 17]). Coupled with a lightweight visualization library, that is capable of handling large amounts of data (i.e., vis.js [1, 2]), as well as exporting data in a .csv to be used by a multitude of third-party applications, TrEcSim provides the possibility of customizing over a dozen different attributes and analyzing the simulated scenarios.

In our dynamic model we start off with a given, well-defined network topology $G = (v, \varepsilon)$ consisting of the set v of nodes/agents, and the ε set of weightless edges, representing relationships. With a fixed network topology, we define a set of attributes that aim at creating the environment needed for transacting various goods between each economic agent; we will use the v_E notation to represent an economic agent, where $v_E \in v$:

- Each v_E economic agent has a unique subset of demands D from the global set of demands D_G: $D \subset D_G = \{D_1, D_2, D_3, \ldots, D_i, \ldots, D_n\}$, where $i \in \mathbf{N}$ and $i = \overline{1, n}$
- A given product Pr fulfills the demand of the economic agent for a given demand D: $Pr \in Pr_G = \{Pr_1, Pr_2, Pr_3, \ldots, Pr_i, \ldots, Pr_n\}$, where $i \in \mathbf{N}$ and $i = \overline{1, n}$
- Product Pr would be needed in a certain quantity Qt: $Qt_{Pr} \in Qt_{Pr_G} = \{Qt_{Pr_1}, Qt_{Pr_2}, Qt_{Pr_3}, \ldots, Qt_{Pr_i}, \ldots, Qt_{Pr_n}\}$, where $i \in \mathbf{N}$ and $i = \overline{1, n}$, and with a given importance I: $I \in I_{Pr} = \{I_{Pr_1}, I_{Pr_2}, I_{Pr_3}, \ldots, I_{Pr_i}, \ldots, I_{Pr_N}\}$, where $i \in \mathbf{N}$ and $i = \overline{1, n}$, both attributes unique for each agent and product
- A Pr product would be chosen over another based on its quality Q: $Q_{Pr} \in Q = \{Q_{Pr_1}, Q_{Pr_2}, Q_{Pr_3}, \ldots, Q_{Pr_i}, \ldots, Q_{Pr_n}\}$, where $i \in \mathbf{N}$ and $i = \overline{1, n}$, while $Q_{Pr_i} \in [0, 100]$
- Each Pr product would have a specific (pricing) value V_{Pr}, which can be defined using Eq. (1) for each product, based on the initial cost of the product, $(V_{Pr_{initial}})$ and the number of Pr products in the network, Qt_{Pr_n}; V_{Pr} can also be defined by a preset, unique probabilistic value: $V_{Pr} \in V_{Pr_G} = \{V_{Pr_1}, V_{Pr_2}, V_{Pr_3}, \ldots, V_{Pr_i}, \ldots, V_{Pr_n}\}$, where $i \in \mathbf{N}$ and $i = \overline{1, n}$, while $V_{Pr_i} \in \mathbf{R}_+$

$$V_{Pr} = \frac{1}{V_{Pr_{initial}} \cdot Qt_{Pr_n}} \tag{1}$$

Each v_E economic agent would have a certain chance of becoming a producer—noted as v_P, where $v_P \in v$—of a subset of products Pr. This means that the agent is able to transact its produced goods in order to obtain currency, either directly—v_P is directly linked to the specific v_E economic agent—or indirectly, via one or more intermediate v_E agents, designated as *middlemen* and noted with v_M, where $v_M \in v$. Producers would also retain the ability to purchase certain products, as these would be used to fulfill their subset of demands D, or be used in production as raw material by end-products.

Generating the specific products would be done taking into consideration the aforementioned attributes for each individual product Pr: Q, Qt and V. These—

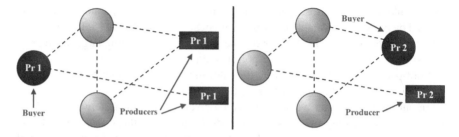

Fig. 1 Visual representation of various user-roles in the economic network and of how roles can change when taking into consideration different products. An economic agent v_E producing Pr_1 (represented with a violet rectangle) may also be a v_E in need of Pr_2—produced by the same agent—while all other agents, who may or may not be middlemen in these two transactions, are represented with a blue circle

along with the total number of products to be generated, Pr_n—would be defined at the point of the creation of a new simulation, either by specifying the number for Pr_n, where $Pr_n \in [0, v_n]$, or by a calculated percentage. Once the total number of products has been determined and created, the simulator proceeds to instantiate the producers.

Defining the producers is done by either explicitly defining their total number— $v_{P_n} \in [0, v_n]$, where v_n represents the total number of economic agents—or their overall percentage. Each v_P would be randomly chosen from the set of agents v_E, but being a producer would not exclude the agent's ability of also being a consumer (refer to Fig. 1 for an illustrative example).

In order to have a constant supply and demand within the network of economic agents, the subset of demands D needs to be defined for each v_E agent, along with the required product $v_{E_{D_{Pr}}}$, where $v_{E_{D_{Pr}}} \in Pr_G$. This product, while defined at the start of the simulation, is not necessarily assigned for production by a v_{E_P}; however, during the simulation, TrEcSim can assign its production to any v_E which would benefit of its production. As such, the demand subset is defined using one of the following options:

- Single product: A given node may or may not require a particular product, regardless if it is currently produced in the current network or not
- Single from production: Similarly to the previous option, but with reassurance that this demanded product indeed exists (is produced) in the present network: $\forall \ v_{E_{D_{Pr}}} \ \exists! \ P_G$
- Multiple products: Defines the probability that a given node may have several demands, which may or may not be fulfilled
- Multiple from production: Not unlike the above option, except that all products are also present in the product-pool: $\forall \ v_{E_{Pr}} \in Pr_G$

The overall simulation duration would be quantified, as a substitute for time in a real-life scenario, by iteration cycles, which in turn consists of two phases for each v_E: a transactional and a decisional phase.

3.1 Transactional Phase

In the transactional phase, all existing economic agents would be able to identify its current subset of demands D, based on their importance factor I, after which they proceed searching for one or more producers v_P (taking into consideration Qt quantity needed, Q quality, and V pricing). Each agent will then be able to choose between settling for a cheaper product, with a lower quality-factor over a more expensive product (with a better quality), even if limited by purchasable quantity.

Whilst the starting-value of $Pr \in Pr_G$ is defined at the start of the simulation, it is subject to change during the decisional phase. Due to the fact that each middleman retains a certain percentage of the final revenue, it is therefore directly influenced by the number of intermediate economic agents in the current stage of the simulation.

Using a breadth-first search, we are able to determine the shortest path to each v_P producer that produces or intermediates product $Pr \in Pr_G$ that is needed by the current agent. As soon as all of the producers are identified, the simulation proceeds at computing all possible ways at meeting the respective agent's demands, prioritizing them based on importance, quantity needed, quality, and available funds. As already mentioned, the final price of the product is determined by its original value plus the profit that is subtracted by each middleman. To this end, in order to establish the final cost of the product Pr (herein including the gained profit φ), we first determine its initial value by applying Eq. (2):

$$Ci_Q(Pr) = Cb(Pr) \cdot (1 + Ip_Q)^Q \tag{2}$$

where $Ci_Q(Pr)$ is the initial cost of producing Pr—taking into consideration its quality factor $Q \in [0, 100]$—$Cb(Pr)$ is the base-cost of product Pr, while Ip_Q is the increased price of the product per each quality value. Consequently, we determine the final cost of the product, namely the price that the respective v_E agent would pay:

$$Cf(Pr) = Ci_Q(Pr) \cdot (1 + Ip_I)^{v_{M_n}} \tag{3}$$

In (3) Ip_I is the increased price per middleman, while v_{M_n} is the total number of intermediaries, $v_{M_n} \in [0, v_n]$. This means that the closer the buyer and the producer are to each other, the less the product will cost, and at the same time the more the producer can retain its profit. Hence, in the particular case in which $n = 0$, meaning that there are no middlemen, the initial cost will be the price with which the buyer will buy the given product:

$$Cf(Pr) = Ci_Q(Pr) \tag{4}$$

In order to determine the final profit φ of the last economic agent v_E in the chain of intermediaries, we have to first determine the subtotal profit $\varphi_{subtotal_{v_{M_n}}}$ earned prior to the current simulation step:

$$\varphi_{subtotal_{v_{M_n}}} = \frac{Cf(T)}{1 + Ip_I} \tag{5}$$

where *Cf(T)* is the final cost of the given transaction *T* which can be determined with (6):

$$Cf(T) = Qt_{Pr} \cdot Cf(Pr) \tag{6}$$

where Qt_{Pr} is the quantity of *Pr* being bought. To this end, the profit of the last middleman in the chain of intermediaries can be determined using the following expression:

$$\varphi_{v_{M_n}} = Cf(T) - \frac{Cf(T)}{1 + Ip_I} \tag{7}$$

Using (7), we express the profit of a given middleman using a more general expression:

$$
\begin{aligned}
\varphi_{v_{M_i}} &= \varphi_{subtotal_{v_{M_{i+1}}}} - \varphi_{subtotal_{v_{M_i}}} \\
&= \varphi_{subtotal_{v_{M_i}}} \cdot (1 + Ip_I) - \varphi_{subtotal_{v_{M_i}}} \\
&= \varphi_{subtotal_{v_{M_i}}} \cdot Ip_I
\end{aligned}
\tag{8}
$$

where $i = \overline{1, n-1}$. The profit of a single intermediary is expressed using (9):

$$\varphi_{subtotal_{v_{M_i}}} = \frac{\varphi_{subtotal_{v_{M_{i+1}}}}}{1 + Ip_I} \tag{9}$$

where $i = \overline{1, n-1}$. Similar to (8), but taking into consideration the subtotal profit of the next middleman in the chain of intermediaries, we express $\varphi_{subtotal_{v_{M_i}}}$ using the following formula:

$$
\begin{aligned}
\varphi_{subtotal_{v_{M_i}}} &= \varphi_{subtotal_{v_{M_{i+1}}}} - \frac{\varphi_{subtotal_{v_{M_i}}}}{1 + Ip_I} \\
&= \varphi_{subtotal_{v_{M_i}}} \cdot \frac{Ip_I}{1 + Ip_I}
\end{aligned}
\tag{10}
$$

Before committing the respective goods, the algorithm would check the following aspects of the potential transaction: required quantity and if it can be supplied by the producer in full or not, as well as the available currency and—in case of insufficient funds—the maximum quantity for the given amount. At the end of the transaction phase, each agent will take turns in initiating the identified transactions, fulfilling their demands for the required products in exchange for a total amount of currency (total value). This value would then be either split into multiple payments based on profit percentage for multiple middlemen, or kept in full by the respective producer. In this last step of the transaction phase, the value distribution (payoff) for each economic agent would be calculated, in order to be used in the next phase.

3.2 Decisional Phase

The transactional phase, while important from an economic point of view, does not entail by itself the network's dynamicity. On the other hand, it is in the decisional phase that actions which alter the network topology take place, namely one of the following actions:

- **Action 1**: Creating new links between two economic agents and thus circumventing a given number of middlemen. Based on information gathered so far, the algorithm computes—for each v_E in the network—which economic agent would be best suited to link to, in order to improve the current agent's economic stance. As already mentioned, the profit that a given producer v_P makes in the transactional phase is directly related to quality, quantity, and value of product Pr; this in turn is greatly influenced by the total number of middlemen (v_{M_n}) involved in the transaction. Such a process can be modeled, at the start of the simulation, using either a probabilistic or deterministically calculated value. This value, however, is greatly influenced by its worth during the simulation process compared to all other actions. Settling for this action would also be suitable if, for instance, the demand for new or improved product is much less than the current (global) set of products: $D_{G_n} \ll Pr_{G_n}$
- **Action 2**: Investing in the creation of a new product by allowing the current economic agent to start producing a specific Pr product, based on ever-growing demands: $D_{Pr_n} \gg Pr_{G_n}$
- **Action 3**: Invest in improving current production quality or quantity by looking through all of economic agent's current products and deciding upon improving either quality or quantity for a given product Pr
- **Action 4**: Expanding the network by creating a new economic agent. This option requires that the algorithm analyze several attributes before computing its outcome: the percentage of the current funds which will be transferred to the new agent; the advantages and disadvantages of creating different products; the advantages and disadvantages of being linked to the new agent; how much of the current agent's debt it should inherit, etc.

3.3 Agent Behavior Algorithm

The aforementioned heuristic algorithm is custom created with specific components in mind, so that based on the described scenario, it is able to choose the best possible solution for each economic agent that it drives. To this end, our algorithm uses the following objects created during initial simulation setup—but updated during each iteration cycle—as data-pool:

- Network topology, including the distinct user-defined settings
- Product-pool
- Producers and data related to their products
- The demanded products for each economic agent

The heuristic algorithm is split into three main components, resembling the before-mentioned phases it drives along with the simulation initialization process. Also, to better convey the function of the algorithm, we describe it by using the flowchart from Fig. 2.

4 Simulation Results

In this section we simulate transactions in economic networks with TrEcSim, by taking into consideration the fundamental network topologies.

Each presented simulation starts off with the exact same settings, and runs the same number of iterations (i.e., 400). The results presented in our charts are obtained by determining the average values of ten independent simulations for each network topology.

Nonetheless, our simulations are not intended to be used for comparing TrEcSim with other approaches, but to obtain qualitative results in terms of payoff distribution.

4.1 Small-World Network Topology

The first network topology that we simulate is a small-world network with 500 v_E agents, 5328 links and 30 Pr product types, with an initial production count of 45 units for each individual product.

In order to better differentiate the behavior of the v_E in the network when limiting certain behavioral aspects and analyzing the results, we simulate two distinct scenarios for all network topologies, namely

- **Scenario 1**: All settings are left at default values, as well as allowing all agent-decisions (Actions 2, 3 and 4) to take place using the same probability ($\approx 33\%$), except the possibility of creating new links between themselves (Action 1); this

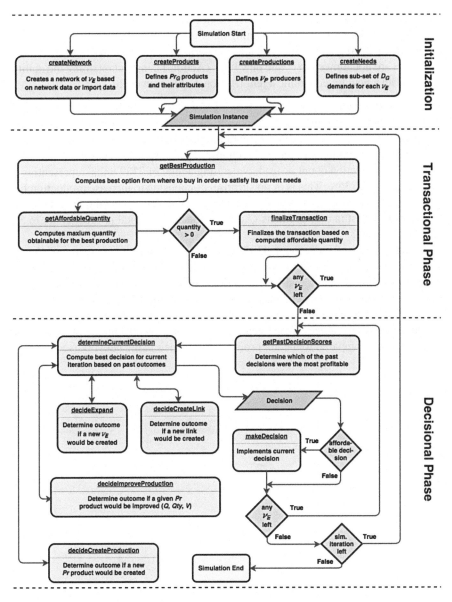

Fig. 2 Flowchart describing the implementation of TrEcSim's methods. During initialization, the simulator creates the network-configuration, products, productions, and needs, based on input-data, creating a simulation instance to be used in the later phases. In the transactional phase, TrEcSim iterates over each existing economic agent, determines the best option for the desired product, and finalizes the respective transaction, should it be possible. In the decisional phase, the algorithm computes the best possible action for the current economic agent to make; this is based not only on past outcomes, but also on affordability, and is done for each agent in the network

is achieved by setting the probability of an economic agent choosing this action to 0%, and normalizing the remaining ones

− **Scenario 2**: Much like the previous scenario, except that this time economic agents are not allowed to create new links (Action 1), instead retaining the ability to create a new v_E economic agent (Action 4) per each cycle

After 400 iterations, using the hypothesis for Scenario 1, the simulation creates 294 (59%) new economic agents, implicitly linking them to the agent choosing the expansion. The economic agents, without the possibility of shortening their path to the buyer (by deciding for Action 1), choose, by a very large margin, to invest in increasing production count instead, generating a total of 2529 units, representing a big increase of 6735%. This is of no surprise, as choosing to increase production for the same cost (Action 3) is the only remaining viable action, leaving only a couple of agents to invest in new products (Action 2), increasing total product count to 52 (a 73% increase).

Using Scenario 2, we count 9335 new links, meaning an increase of 175%; this result suggests that our custom algorithm behind TrEcSim—just as we would expect it to happen in any real-life scenario—deems it more advantageous from an economic standpoint to extend each producer's reach in the network (deciding for Action 1), therefore gaining several advantages within the process.

One of these advantages is the retention of an ever-growing amount of income for the producers, compared to the net profit of middlemen in the network (see Fig. 3).

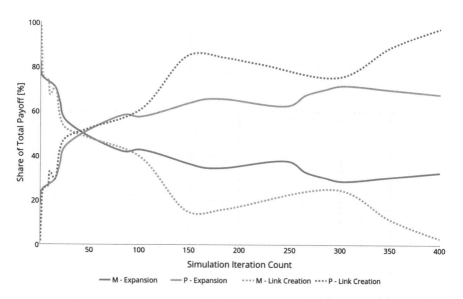

Fig. 3 Share of total payoff as a function of iteration count in a small-world network for producers (P) and middlemen (M), allowing economic agents to choose either expansion (continuous) or link creation (dotted)

Another advantage when agents are allowed to invest in creating new links (by deciding for Action 1) is the ability to sell the same amount of products cheaper, or—even better from an economic point of view—to sell more products for the same amount of production costs. To this end, production increases to 92 units, representing an increase of 38% for each product, supplemented by an increase of 9 new products being produced (30%).

In both cases, we observe, though at a different rate, the ever-growing payoff for each economic agent even producing a given product, compared to the middlemen in the current network. These middlemen, who, initially took advantage from their topocratic position, could not keep up in the end with the economic freedom which allows producers to freely invest their revenue in either new economic links or a more efficient production.

4.2 Random Network Topology

In order to simulate a random network, we create a graph with 1000 v_E agents and starting with the exact same conditions as previously: economic agents producing 30 Pr products (45 units each) in a network with 80,776 links.

After simulating Scenario 1, the number of products increases to 52 (an effective increase of 73%), while the total number of production units increases to 1409 (an estimate increase of 3031%). Similar to the simulation corresponding to the small-world topology, the only viable option the agents have is to invest in either Action 2 or Action 3, compared to the benefit a new link would bring by choosing Action 1. Nonetheless, producers eventually start obtaining more after each transaction compared to middlemen, adopting an effective strategy against a topocratic, oligarchic society.

For the second scenario, much like in the previous scenario, TrEcSim slowly but consistently increases the number of links in the network to 104,072,—meaning an effective increase of 29%—obtaining the dotted traces in Fig. 4.

The similarity between Figs. 3 and 4 can be interpreted as follows: having almost the same average degree as any random node in a small-world network, economic agents in a random network take advantage of their increased production quality or lower prices per each unit of production, as opposed to middlemen, who have no other choice but to increase the price for each product intermediated or risk ending up without income. To this end, production of individual units has also increased in number, namely to 89 units (an increase of 98%).

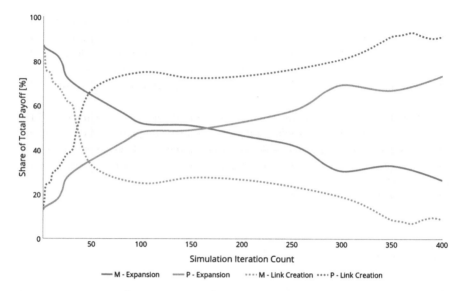

Fig. 4 Share of total payoff as a function of iteration count in a random network for producers (P) and middlemen (M), allowing economic agents to choose either expansion (continuous) or link creation (dotted)

4.3 Scale-Free Network Topology

The most interesting result from all simulations pertains to the scale-free topology. In this case, we start off with 1000 v_E agents, 3092 links, 30 types of Pr products, and an initial production of 45 units of each created good.

In this case, the evolution of the payoff for the producers and middlemen is of more importance, due to the fact that the individual economic agents of a scale-free network are linked to so-called hubs—rather than among themselves—who, as well-positioned middlemen, take advantage of their topocratic position within the network. This is the main reason why the traces in Fig. 5 for both scenarios evolve at a much slower rate, compared to the previous simulations. Not only this, but the payoff is also a lot more sporadic, further supporting the assumptions affirming that the producers impose themselves with difficulty due to their initial topocratic disadvantage, compared to other network topologies.

After applying the Scenario 1, the number of products increases to 34 units, meaning an estimative increase of 13%, while individual product count increased to 2085 units (an increase of 4533%). The development of this simulation yields an interesting observation, namely that even after the 400 cycles of simulation the producers do not manage to rival the income of the middlemen, strongly supporting the hypothesis that the clustering is still present in the network, keeping the remaining middlemen in favorable positions.

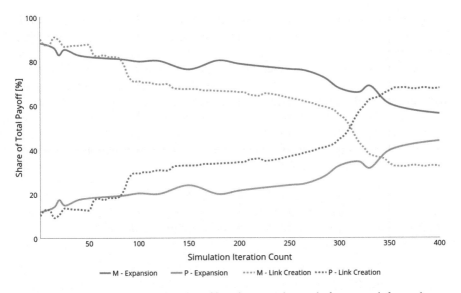

Fig. 5 Share of total payoff as a function of iteration count in a scale-free network for producers (P) and middlemen (M), allowing economic agents to choose either expansion (continuous) or link creation (dotted)

In Scenario 2, where economic agents are allowed to create new links, the number of links increases to 19,224, meaning an effective increase of 522%, representing the preferred choice of investment, if allowed. Only when they manage to create enough new links within the network, we begin to see an increased income for the producers, in comparison to the middlemen. To this end, in this scenario, we observe only a relatively small increase of products (14 new products, roughly equivalent to 47%), and with a final production count of 73 units for each product, representing an increase of 62%. Much like in the case of the previous two network topologies, agents do not have the option to expand, thus providing incentives for increasing the number of economic agents in the network.

5 Conclusions

5.1 Discussion

Complex networks are comprehensively studied due to their important applications in various fields, from medicine and sociology, to architecture, engineering, and economics; as such, they can also be applied to an amalgam of these fields. Nowadays, economic and social networks are mostly used for helping us understand our society, and the role we all play in it, especially from a socioeconomic point of view. For instance, we would often identify individuals who would benefit from a topological opportunity, though without any creative contribution to the network itself. Hence, we cannot fully understand a meritocratic network without factoring in topocracy.

TrEcSim was specifically created to simulate various economic activities, validating—or, in some cases, even invalidating—theories, models, and assumptions in the process. One of these models is Borondo's mathematical—yet static—model, in which they came to the conclusion that with recent changes to our society, meritocracy would quickly gain advantage over a long chain of intermediation in a poorly connected network [7]. This statement is also confirmed by us after simulating multiple network topologies, concluding that regardless of the network's topology, using relevant data as a starting-point and allowing producers to freely invest their revenue in new economic links, additional v_E agents, or in efficient production, the network always shifts toward a meritocratic layout—not unlike our own, well-established network, of which we are all a part of.

5.2 Future Work

Extra effort will be put into the improvement of the TrEcSim's heuristic algorithm. As such, the algorithm will be enhanced to the point at which it will be able not only to look back at past decisions made and analyze its outcome, but to use this information and create a buffer simulation; in other words, a side-simulation, of several steps ahead while probabilistically analyze its results, thus greatly improving the accuracy of choices. This, along with an improved implementation of two main economic theories – "the theory of marginality" and "the labor theory of values" – will greatly contribute to TrEcSim's applicability.

At the same time, we aim at further improving the realism of our model by taking into consideration information asymmetry, which often occurs in transactions, as well the role of government involvement and regulatory red-taping.

Acknowledgements The authors would like to the thank Alexandru Stana for his contributions to the development of TrEcSim's first version. We also want to express our gratitude to both Alexandru Topirceanu and Alexandru Iovanovici for their insights and constant support throughout the development phases of our simulator.

References

1. Almende, B.V.: vis.js - a dynamic, browser based visualization library. http://visjs.org/ (2016)
2. Almende, B.V., Thieurmel, B.: visnetwork: network visualization using 'vis.js' library. https://cran.r-project.org/web/packages/visNetwork/index.html (2016), r package version 0.2 1
3. Arthur, W.B.: Complexity and the Economy. Oxford University Press, New York (2015)
4. Avila, R., Broed, R., Pereira, A.: Ecolego-a toolbox for radioecological risk assessment. No. IAEA-CN–109 (VER. 3) (2003)

5. Bastian, M., Heymann, S., Jacomy, M.: Gephi: an open source software for exploring and manipulating networks. In: Proceedings of the 3rd International Conference on Weblogs and Social Media, pp. 361–362. AAAI, Menlo Park (2009)
6. Blume, L.E., Durlauf, S.N.E.: The Economy as an Evolving Complex System III: Current Perspectives and Future Directions. Oxford University Press, New York (2005)
7. Borondo, J., Borondo, F., Rodriguez-Sickert, C., Hidalgo, C.A.: To each according to its degree: the meritocracy and topocracy of embedded markets. Sci. Rep. **4**, 3784 (2014)
8. Broed, R., Xu, S.: User's Manual for Ecolego Toolbox and the Discretization Block. Swedish Radiation Protection Authority, no. SSI–2008-10. edn. (2008)
9. Chatterjee, A., Chakrabarti, B.K.E.: Econophysics of Markets and Business Networks. Springer, Berlin (2007)
10. Dawid, H., Gemkow, S., Harting, P., van der Hoog, S., Neugart, M.: Agent-based macroeconomic modeling and policy analysis: the eurace@ unibi model. In: Chen, S.H., Kaboudan, M. (eds.) Handbook on Computational Economics and Finance. Oxford University Press, Oxford (2014)
11. Deissenberg, C., van der Hoog, S., Dawid, H.: EURACE: a massively parallel agent-based model of the european economy. Appl. Math. Comput. **204**(2), 541–552 (2008)
12. Duval, J.S.: Version 3.0 of eminers-economic mineral resource simulator. US Gelogical Survey No. 2004–1344 (2012)
13. Easley, D., Kleinberg, J.: Networks Crowds, and Markets, vol. 8. Cambridge University Press, Cambridge (2010)
14. Gartman, D.: Marx and the labor process: an interpretation. Insurg. Soc. **8**(2–3), 97–108 (1978)
15. Heer, J., Boyd, D.: Vizster: visualizing online social networks. In: Proceedings of the 2005 IEEE Symposium on Information Visualization (INFOVIS), pp. 32–39. IEEE, New York (2005)
16. Jackson, M.O.: Social and Economic Networks. Princeton University Press, Princeton (2008)
17. Jacomy, M., Venturini, T., Heymann, S., Bastian, M.: Forceatlas2, a continuous graph layout algorithm for handy network visualization designed for the gephi software. PLoS One **9**(6), e98679 (2014)
18. Keen, S.: Finance and economic breakdown: modeling minsky's 'financial instability hypothesis'. J. Post Keynesian Econ. **17**(4), 607–635 (1995)
19. Marshall, A.: Principles of Economics: Unabridged Eighth Edition. Cosimo, New York (2009)
20. Marx, K.: Value, Price and Profit. International Publishers, New York (1865/1969)
21. Menger, K.: On the origin of money. Econ. J. **2**(6), 239–255 (1892)
22. Newman, M.E.: The structure and and function of complex networks. SIAM Rev. **45**(2), 167–256 (2003)
23. Topirceanu, A., Barina, G., Udrescu, M.: Musenet: collaboration in the music artists industry. In: Proceedings of the 1st European Network Intelligence Conference (ENIC). IEEE, New York (2014)
24. Xiang, L., Yu, Y.J., Guanrong, C.: On the topology of the world exchange arrangements web. Phys. A Stat. Mech. Appl. **343**, 573–582 (2004)

Towards an ILP Approach for Learning Privacy Heuristics from Users' Regrets

Nicolás Emilio Díaz Ferreyra, Rene Meis, and Maritta Heisel

Abstract Disclosing private information in Social Network Sites (SNSs) often results in unwanted incidents for the users (such as bad image, identity theft, or unjustified discrimination), along with a feeling of regret and repentance. Regrettable online self-disclosure experiences can be seen as sources of privacy heuristics (best practices) that can help shaping better privacy awareness mechanisms. Considering deleted posts as an explicit manifestation of users' regrets, we propose an Inductive Logic Programming (ILP) approach for learning privacy heuristics. In this paper we introduce the motivating scenario and the theoretical foundations of this approach, and we provide an initial assessment towards its implementation.

Keywords Adaptive privacy · Self-disclosure · Awareness · Social network sites · Inductive logic programming

1 Introduction

Privacy norms play an important role in the construction of people's identity at an individual and collective level [3]. However, users of Social Network Sites (SNSs) like Facebook or Twitter seem to deliberately violate their privacy norms while interacting within these media platforms [2]. That is, users often share private or sensitive information that reaches unintended audiences in many cases. Consequently, users sharing acts draw unwanted incidents (like stalking, identity theft, sextortion, and job loss), along with a feeling of regret [5].

As one can see, there are two central aspects of information sharing: (1) Not every information is appropriate to be mentioned in a particular context, and (2) the

N. E. Díaz Ferreyra (✉) · R. Meis · M. Heisel
University of Duisburg Essen, Duisburg, Germany
e-mail: nicolas.diaz-ferreyra@uni-due.de; rene.meis@uni-due.de; maritta.heisel@uni-due.de
https://www.ucsm.info/

© Springer International Publishing AG, part of Springer Nature 2018   187
R. Alhajj et al. (eds.), *Network Intelligence Meets User Centered Social Media Networks*, Lecture Notes in Social Networks,
https://doi.org/10.1007/978-3-319-90312-5_13

audience to which information is disclosed influences such context. For instance, a user can consider acceptable to share his/her political affiliation with his family, but inappropriate with his/her work colleagues. Such aspects, which have been already identified in privacy theories (e.g., "contextual integrity" by Nissenbaum [10]), are basic for the definition of privacy norms and consequently for individual privacy control.

In order to create a more privacy-aware social environment, media technologies should support the users by providing guidance and feedback on their disclosures. Moreover, awareness mechanisms should be able to identify and learn from regrettable experiences in order to provide effective privacy support. In line with these premises, Diaz Ferreyra et al. introduced the concept of Instructional Awareness System (IAS) which uses a privacy heuristics database in order to generate adaptive awareness [4, 5]. In this paper, we propose to take a closer look into users' deleted posts and consider them as potential sources of privacy heuristics. Using Inductive Logic Programming (ILP) as the learning approach and deleted posts as the training set, we provide an initial assessment towards a privacy heuristics inference engine for IAS.

The content of this paper is organized as follows. In the next section, we discuss related work. Section 3 introduces the basic notions of ILP that form the paper's background. Following, in Sect. 4, we discuss how regrets can be identified and retrieved from users' deleted posts. In Sect. 5, we introduce a representation for regrettable scenarios and the initial components of an ILP system for learning privacy heuristics. Finally, we conclude in Sect. 6 with an outline and directions for future work.

2 Related Work

In the context of SNSs, adaptive privacy awareness systems seek to provide tailored feedback to the users when they attempt to reveal private information inside a post or in their profiles. Such feedback normally takes the form of a warning message that is displayed when a potentially unwanted disclosure is detected (e.g., a user revealing his/her bank account details in a public post on Facebook). For instance, Caliki et al. [2] developed a system which learns privacy rules from the user's previous sharing history, to use them later on as a criterion for raising awareness. That is, under the assumption that the user has never revealed private content to an unintended audience, the system infers *allow/deny(data, audience)* privacy rules through a machine learning engine (where "data" is a piece of private information like *bank account*, and "audience" is a sub-group of the user's Facebook friends). If a rule derived during the learning stage of the system is later violated in its operational stage, then a warning message is displayed. In line with this approach, Fang and LeFevre [6] introduced a system which analyses the information disclosed inside the user's Facebook profile in order to derive privacy preferences. This system recommends which attributes of the profile should be visible or not within a certain group of friends.

Following similar adaptivity principles, Díaz Ferreyra et al. developed an architecture which prescribes the basic components of an Instructional Awareness System (IAS) [4]. Using privacy best practices stored inside a Privacy Heuristics Data Base (PHDB), IAS generates a personalized warning message when a user attempts to reveal private data in a SNS post. The privacy heuristics in the PHDB are the outcome of a method which analyses the risks of regrettable online experiences reported by the users. That is, the input of the method are the experiences that users have reported themselves (to the development team of IAS, for instance), or the outcome of an empirical research (e.g., questionnaires or face-to-face interviews). This approach is effective for building a baseline of heuristics prior to the execution of the system. However, eliciting new entries of the PHDB requires the execution of this process which can be expensive and inefficient in terms of the resources and time needed to conduct interviews and process their outcome. Therefore, a *run-time* (*online*) approach for learning privacy heuristics would be beneficial for keeping up to date the content of IAS's PHDB.

3 Theoretical Background

This work proposes to endow IAS with a Privacy Heuristics Learning Engine (PHLE) based in principles of ILP. In this section we introduce the theoretical foundations of ILP and discuss its potential for carrying forward this task.

3.1 The ILP Problem

ILP is a discipline which employs techniques from machine learning and logic programming to infer hypotheses H from a set of observations and some background knowledge B [8]. Such observations are a set of positive and negative examples $E = E^+ \cup E^-$ of a concept to learn, and like the background knowledge, they are expressed as logic programs. The goal of ILP is to generate a logic program where positive examples are satisfied and negative examples are not. Let us consider that E^+ is expressed as ground unit definite clauses and E^- as ground unit headless Horn clauses of a single target predicate [7].[1] Then E contains ground unit clauses of a single predicate, and we can specify the general ILP problem as $B \wedge H \models E$. Since the goal is to find the simplest hypothesis, each clause in H should explain at least one example. If we consider H and E as single Horn clauses, then this expression can be rearranged as $B \wedge \neg E \models \neg H$.

Let $\neg\bot$ be the (potentially infinite) conjunction of ground literals which are true in all models of $B \wedge \neg E$ [9]. Considering that $\neg H$ must be true in every model of $B \wedge \neg E$, then it must contain a subset of the ground literals in $\neg\bot$. That is, $B \wedge \neg E \models$

---

[1]Let $e(X)$ be the predicate which defines the examples, and $L = L_1, \ldots, L_n$ a set of ground literals which subsume the variable X. Then, positive examples can be expressed as $e(L_i).$, and negative examples as : $-e(L_j), \forall\, 1 \leq i, j \leq n$.

$\neg\perp$ and $\neg\perp \vDash \neg H$. Rearranging this expression, we obtain $H \vDash \perp$, which means that a subset of the solutions for H can be derived from $\perp$ by generalizing each given example in E. This means, $\perp$ is a clause that θ-subsumes[2] each example $e \in E$ [9]. This way, the final $\perp$ can be constructed as the disjunction of the body literals of all these derived clauses [9]. However, since $\perp$ can have infinite cardinality, the search space of those clauses which imply $\perp$ can also be infinite. In order to bound the search space of consistent and complete hypotheses, $\perp$ is built and generalized using *mode declarations* [9, 11].

3.2 Mode Declarations and Types

Mode declarations are *language bias* (i.e., syntactic restrictions) that are imposed on candidate hypotheses in order to make the search space more efficient [9]. A mode declaration has either the form $modeh(n, s)$ or $modeb(n, s)$ for the head or body of a rule, respectively[9]. Consequently, it restricts the predicates which can occur in the head and body of a rule. The schema s is a ground literal with placemarkers of the form '$+type$', '$-type$', and '$\#type$' standing for input variables, output variables, and constants, respectively [9]. The value n is the *recall* of the mode declarations, and is used to bound the number of times the scheme can be used. Let us consider the following example:

```
:- modeh(1, fly(+bird))?
:- modeb(1, wings(+bird,#property,-int))?
property(has_flight_feathers).
```

The first mode declaration says that general rules may have heads containing the predicate $fly(X)$ where X is a type of $bird$ [1]. The second says that general rules may have bodies containing the predicate $wings(X, has\,flight\,feathers, Y)$, where X is a type of $bird$, Y an $integer$, and $has\,flight\,feathers$ a $property$ [1]. The value of the recall here is one; therefore, both mode declarations can be used once in the construction of the bottom clause.

4   Regrets Identification in SNSs

As discussed in the previous section, ILP is a supervised machine learning approach that can serve the purpose of developing a Privacy Heuristics Learning Engine (PHLE) for IAS. One sub-task in the development of such PHLE is to generate a set of examples (training set) to provide as input to the hypotheses derivation process. Since in many cases the information enclosed within a post has a private or sensitive semantics, posts become elemental units for user-centered privacy analysis in SNSs. In this sense, performing a privacy analysis of a post consists of determining

---

[2]A clause c_1 θ-subsumes a clause c_2 if and only if there exists a substitution θ such that $c_1\theta \subseteq c_2$. Consequently c_1 is a generalization of c_2 (and c_2 specialization of c_1) under θ-subsumption [8].

> **USER'S POST** *"What a lame environment at the office...I can only hear people complain! Thanks God it's Friday! #tgif"*
>
> ---
>
> **Actual Audience:** PUBLIC.
> **Unintended Audience:** The user's work colleagues, or superior.
> **Unwanted Incidents:** Wake-up call from superior; bad image; job loss.

Fig. 1 Example of self-disclosure scenario

the existence of private information enclosed in it. Such privacy analysis of user-generated content requires a taxonomy of attributes that can be considered as private. For this purpose, Díaz Ferreyra et al. proposed to organize the users' private or sensitive personal attributes around different high-level categories called the "self-disclosure dimensions" (i.e., demographics, sexual profile, political attitudes, religious beliefs, health factors and condition, location, administrative, contact, and sentiment) [5]. Each dimension consists of a set of Surveillance Attributes (SAs)[3] which allow analyzing from a user-centered perspective the information disclosed inside a post. For instance, in the scenario described in Fig. 1, the SAs *employment status*, *work location*, and *negative* sentiment are disclosed inside a post.

When one or more SAs reach an unintended audience, an unwanted incident can take place and result in a feeling of regret. In Fig. 1, the post reached the user's work acquaintances causing a wake-up call from superior, together with a negative impact on the user's image, and eventual job loss. Since the likelihood and impact of such incidents define the risks of the given scenario, they can be used to model a regret. That is, regrets can be expressed in terms of the SAs, unintended audience, and the associated risks (i.e., likelihood and impact of an unwanted incident). Risks, likelihood and consequence, can be expressed on a nominal scale. In this sense, a consequence is a value on an impact scale such as insignificant, minor, moderate, major, or catastrophic. Likewise, the likelihood is a value on a frequency scale such as rare, unlikely, possible, likely, and certain. Finally, a risk is a value obtained from the likelihood and consequence of the unwanted incident and expressed on a scale such as very low, low, high, and very high.

5   Towards the Development of a Privacy Heuristics Learning Engine

So far we have discussed the concepts that serve in the identification of regrettable posts and therefore for creating examples for a PHLE. However, in practice, when a post is deleted there is not much information about the risks and, moreover, if the

---

[3]SAs are those which can be linked to an individual, groups or communities and can raise privacy concerns related to data aggregation, probabilistic re-identification and undesirable social categorizations [5].

deleted post has resulted or not in a regret for the user. In this section we provide an approach on how this missing information can be retrieved and used later on for the development of privacy heuristics. We also introduce a syntax for such heuristics in the IAS's PHDB together with the respective mode declarations for their automatic inference.

5.1 Privacy Heuristics

Snippet 1 describes a syntax for the entries in the PHDB, namely the hypotheses (i.e., regret conditions) to be learned by the PHLE. These learned hypotheses are the privacy heuristics which will help to identify potential regrettable scenarios in the future. That is, when a user attempts to disclose SAs inside a post, IAS should query the PHDB in order to verify that this will not lead to a potential regret. In other words, we want to find the consequence (Cons) and frequency (Freq) of a potential unwanted incident (Unwi) based on the disclosed SAs ([X|Xs]) and the list of users that conform the post's audience ([Y|Ys]). The *regret* predicate models this conceptual relation, and is the entry point for querying the PHDB. Consequently, [X|Xs] and [Y|Ys] are input variables of a *regret-?* query, and Unwi, Cons, and Freq are the output variables.

The impact of an unwanted incident (and consequently the acceptance level of the risk associated with it) are perceived differently among individuals. For instance, some users might consider the consequence associated with a "wake up call from supervisor" as Insignificant, and others as Catastrophic. This is because individuals have different privacy attitudes which influence the perception level of risky events [5]. In order to model this, we will adopt the Westin's Privacy Index for the classification of the users' privacy attitudes [12]. Such an index classifies individuals into three privacy groups: fundamentalists, pragmatists, and the unconcerned (each group with high, medium, and low levels of privacy concerns, respectively). The predicate *acceptable* represents this relation between the user's privacy attitude and the risk acceptance level associated with it.

```
/*Privacy Heuristic*/
regret([X|Xs], [Y|Ys], Unwi, Cons, Freq):-
   srv_att_list([X|Xs]), usr_list([Y|Ys]), inSIG([Y|Ys], SIG1),
   unwanted_inc([emp_status, work_location, neg], SIG1,
      wake_up_call),
   subset([X|Xs], [emp_status, work_location, neg]),
   risk(wake_up_call, Cons, Freq, Level), not acceptable
   (Att,Level).
```

Snippet 1 Example of a privacy heuristic

Another element of a regret is the audience towards which a set of SAs has been disclosed. Normally in SNSs like Facebook, a user has a list of "friends" composed by other users of Facebook. We will adopt this approach and assume that the user's friends list is grouped into different circles which are constructed under the premises of Social Identity Theory (SIT). Basically SIT postulates that people belong to multiple social identities (for instance, being Swedish, being an athlete, or being a researcher are all examples of identities/identity groups). Since users frequently have a mental conceptualization of the different social identity groups with whom they interact, we will assume that the friends list of a user is clustered into a set of identity groups as suggested by SIT. We can imagine, for instance, groups like *work*, *church*, *gym*, or *choir* depending on the social circles that a user belongs to. The predicate *inSIG* evaluates if the audience to which the post has been disclosed corresponds to one of the social identity groups (SIG) in which the user's friends list is clustered.

To learn the circumstances of a regrettable scenario means to define instances of some of the variables which appear in the body of the *regret* predicate. Particularly, we are interested in knowing which concrete SAs lead to a certain unwanted incident when disclosed to a particular SIG. Consequently, the SAs, unwanted incident and SIG involved in the scenario of Fig. 1, are represented by the ground literals *[emp_status, work_location, neg]*, *wake_up_call*, and *SIG1* respectively (we will consider *SIG1* as the identifier of the SIG "work place"). The relation between these elements is modeled by the *unwanted_inc* predicate, and the *subset* predicate verifies if the SAs involved in the unwanted incident are a subset of the SAs disclosed inside the deleted post. Likewise, the *risk* predicate models the semantic relation between the unwanted incident (*wake_up_call* in this case), its frequency, and its consequence (as described in Sect. 4). As already mentioned, the risk of an unwanted incident (which is expressed as a level on a risk scale) can be acceptable or nor depending on the user's privacy attitude. For instance, we can assume that for fundamentalists only very low risks are acceptable, for a pragmatist very low and low risks are acceptable, and for an unconcerned very low, low, and high risks are acceptable. The acceptance of the risk level will at the end determinate if the disclosure scenario which is being evaluated will lead to a potential state of regret or not.

5.2 Regrets Retrieval

Once the relevant SAs are identified, the information about the causes which motivated the user to delete the post have to be retrieved. Since this information is not evident at a glance, we propose to build an interface which asks questions to the user when a "delete" event takes place. That is, when a post is deleted, we first analyze if it contains SAs, and if so we ask the user for extra information to complete the description of the regrettable scenario. A mock-up of the described interface is depicted in Fig. 2, where the reported risk information corresponds to the post described in Fig. 1.

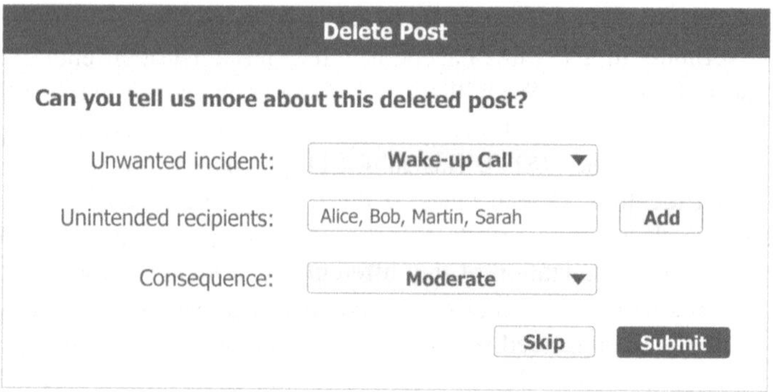

Fig. 2 "Delete Post" interface

Among the information requested from the user, there is the "unintended recipients" of the post. This corresponds to a list of users which were part of the original audience of the post but should not have been included for privacy reasons. In the example, the user reports that the post reached his/her work colleagues Alice, Bob, Martin, and Sarah and has selected them as the unintended recipients. Likewise, the user reports that the unwanted incident has been a *wake-up call* and that the consequence has been perceived by him/her as *moderate*. With this information submitted by the user (i.e., unintended recipients, unwanted incident, and consequence), and the list of SAs extracted from the post, we can create a *regret* predicate consisting of the ground literals: *[emp_status, work_location, neg]*, *[Alice, Bob, Martin, Sarah]*, *wake_up_call*, *moderate*, *certain*. This *regret* predicate represents a concrete regrettable scenario, and is therefore a training example for the PHLE.

Once the regret scenario is submitted by the user, IAS will first store the information about the risks. That is, it will generate a *risk* entry in the PHDB with the information about the unwanted incident and its consequence. Since a *risk* predicate contains also the frequency of the unwanted incident and the risk level, such values must also be generated. For the frequency, we will adopt "certain" for every case since we can assume that if a post containing SAs has been deleted it is because a regret took place. Likewise, risks levels can be defined for every value combination of likelihood and consequence. We will assume that when the likelihood is certain and the consequence is moderate, the risk level is very high. Consequently, for our example, a *risk* predicate with the ground literals *wake_up_call*, *moderate*, *certain*, and *very_high* is created as a new entry in the PHDB. Risks together with the information of their acceptance level (the *acceptable* predicates) are part of the background knowledge of the PHDB. In this case we will consider that the user is a *pragmatist*, and therefore only risks which are *low* or *very_low* are acceptable (see Snippet 2).

```
/*Regret Example*/
regret([emp_status, work_location, neg], [Alice, Bob, Martin,
    Sarah], wake_up_call, moderate, certain).
/*Background Knowledge*/
risk(wake_up_call, moderate, certain, very_high).
sig([Bill, Bob, Sam, Sarah, Alice, John, Martin], SIG1).
acceptable(pragmatist, low).
acceptable(pragmatist, very_low).
```

Snippet 2 A regret example submitted by the user

Another element which is part of the background knowledge are the different SIGs in which the user's friend list is grouped. As can be observed in Snippet 2, the predicate *sig* assigns an identifier to each of the SIGs. For instance, SIG1 refers to a group of users consisting of *[Bill, Bob, Sam, Sarah, Alice, John, Martin]*. This information allows deriving heuristics where the audience is generalized to a SIG. This is, based on the examples of regrettable scenarios, identify the SIG to which a set of SAs should not be revealed to. In Snippet 2, we can see that the unintended recipients could be generalized to SIG1. Consequently, the next time the user attempts to disclose the SAs *[emp_status, work_location, neg]* to an audience containing a member of SIG1, IAS will raise a warning message.

5.3 Mode Declarations and Type Definitions

In order to define the mode declarations, first it is necessary to define the *types* of the elements which are part of them. That is to describe the categories of objects (number, lists, names, etc.) in the domain being modeled. In our case we need the types *Cons, Freq, Unwi, Usr, SA, Att, Level, srv_att_list, usr_list*. Due to space limitations, we only provide a partial description of these in Snippet 3. The missing types can be defined in an analogous way. Once these types are defined, we can then proceed with the definition of the head and body mode declarations. The head of a heuristic (i.e., a *regret* predicate) is a function of objects of the types *srv_att_list, usr_list, Unwi, Cons*, and *Freq*. All of these objects are variables in the head of the *regret* predicate, however (as mentioned in Sect. 5.1), only *srv_att_list* and *usr_list* are input variables. Therefore, the placemarkers of the *modeh* for the head of the *regret* predicate will be +*srv_att_list*, +*usr_list*, -*Unwi*, -*Cons*, and -*Freq*.

```
/*Types*/
SA(emp_status). SA(work_location). SA(neg).
...
srv_att_list([]).
srv_att_list([X|Xs]):-SA(X), srv_att_list(Xs).
/*Mode declarations*/
:- modeh(1, regret(+srv_att_list, +usr_list, -Unwi, -Cons,
     -Freq))?
:- modeb(1, inSIG,(+usr_list, #SIG)?
:- modeb(1, unwanted_inc(#srv_att_list, #SIG, #Unwi))?
:- modeb(1, subset(+srv_att_list, #srv_att_list)?
:- modeb(1, risk(#Unwi, -Cons, -Freq, -Level))?
:- modeb(1, not acceptable(+Att, +Level))?
```

Snippet 3 Types and mode declarations

The body of the *regret* predicate is defined by the predicates *inSIG*, *unwanted_inc*, *subset*, *risk*, and *not acceptable*. Therefore, we need to define a body mode declaration *modeb* for each one of these predicates. Following the same criterion used to define the input and output variables of *modeh*, we define the arguments for each *modeb*. Since many of the predicates used in the body of *regret* are facts or contain grounded literals in their declaration, we express their respective *modeb* using the *#type* placemarker. Such is the case of the *risk* clause which in the body of the *regret* rule has its parameter *Unwi* defined as a constant. Likewise, the argument *SIG* in *inSIG*, *srv_att_list* in *subset*, and all the arguments of *unwanted_inc* are grounded literals in the body of *regret*. Therefore, we express them as constants in their respective body mode declarations.

6   Conclusion and Future Work

So far, we have introduced an ILP model of a PHLE which learns privacy heuristics from regrettable posts in SNSs. We believe that this is a promising approach which will contribute in shaping better and more user-centered preventative technologies. It is a matter of future research to develop a prototype of the model here introduced in order to measure its performance, as to evaluate if adaptive awareness approaches like IAS indeed help to achieve better engagement levels. Other directions for future research involve the development of alternative approaches to Westin's Privacy Index for measuring the users' perceived severity of unwanted incidents (as suggested by Díaz Ferreyra et al. [5]). This involves the development of a user interface (in the form of a short questionnaire) to capture the users' attitudes toward privacy risks in order to improve the performance of the PHDB and consequently of IAS.

Acknowledgements This work was supported by the Deutsche Forschungsgemeinschaft (DFG) under grant No. GRK 2167, Research Training Group "User-Centred Social Media".

References

1. Athakravi, D., Broda, K., Russo, A.: Predicate invention in inductive logic programming. In: OASIcs-OpenAccess Series in Informatics, vol. 28. Schloss Dagstuhl-Leibniz-Zentrum fuer Informatik (2012)
2. Calikli, G., Law, M., Bandara, A.K., Russo, A., Dickens, L., Price, B.A., Stuart, A., Levine, M., Nuseibeh, B.: Privacy dynamics: learning privacy norms for social software. In: Proceedings of the 11th International Symposium on Software Engineering for Adaptive and Self-managing Systems, May 2016, pp. 47–56. ACM, New York (2016)
3. Diaz, C., Gürses, S.: Understanding the landscape of privacy technologies (extended abstract). In: Proceedings of the Information Security Summit, ISS, May 2012, pp. 58–63 (2012)
4. Díaz Ferreyra, N.E., Schäwel, J., Heisel, M., Meske, C.: Addressing self-disclosure in social media: an instructional awareness approach. In: Proceedings of the 2nd ACS/IEEE International Workshop on Online Social Networks Technologies (OSNT), December 2016. ACS/IEEE, Washington/Piscataway (2016)
5. Diaz Ferreyra, N. E., Meis, R., Heisel, M.: Online self-disclosure: from users' regrets to instructional awareness. In: International Cross-Domain Conference for Machine Learning and Knowledge Extraction. Springer, 2017, pp. 83–102
   Díaz Ferreyra, N.E., Meis, R., Heisel, M.: Online self-disclosure: from users' regrets to instructional awareness. In: Proceedings of the IFIP International Cross-Domain Conference (CD-MAKE) (August 2017) (accepted for publication)
6. Fang, L., LeFevre, K.: Privacy wizards for social networking sites. In: Proceedings of the 19th International Conference on World Wide Web, WWW '10, pp. 351–360. ACM, New York (2010). http://doi.acm.org/10.1145/1772690.1772727
7. Muggleton, S.: Inverse entailment and progol. New Gener. Comput. 13(3), 245–286 (1995)
8. Muggleton, S., De Raedt, L.: Inductive logic programming: theory and methods. J. Log. Program. 19, 629–679 (1994)
9. Muggleton, S.H., Firth, J.: CProgol4.4: a tutorial introduction. In: Dzeroski, S., Lavrac, N. (eds.) Relational Data Mining, pp. 160–188, Springer, Berlin (2001). http://www.doc.ic.ac. uk/~shm/Papers/progtuttheo.pdf
10. Nissenbaum, H.: Privacy as contextual integrity. Wash. L. Rev. 79, 119 (2004)
11. Roberts, S.: An Introduction to Progol. Department of Computer Science, University of York, Heslington, York (1997)
12. Westin, A.F.: Privacy and freedom. Wash. Lee L. Rev. 25(1), 166 (1968)

Part V
Content Analysis

Trump Versus Clinton: Twitter Communication During the US Primaries

Jennifer Fromm, Stefanie Melzer, Björn Ross, and Stefan Stieglitz

Abstract When Donald Trump won the Republican nomination and subsequently beat Hillary Clinton in the presidential elections, his success came as a surprise to most observers. This research contributes to understanding the dynamics of this unusual campaign, in which social media played a prominent role. We collected 6099 tweets by both nominees during the presidential primaries and identified the 21 most frequently discussed issues through computer-assisted content analysis. Secondly, we used time series analysis to investigate whether the candidates influenced each other's political agendas. Most tweets by the candidates were found not to be about policy but about parties, other politicians, and the media. Of the political issues that were discussed, the most prominent ones were employment, family, minorities and terrorism. For tweets about minorities, we found possible evidence of agenda setting. We conclude that social media are mainly being used to reach out to supporters, instead of interacting with the opponent.

Keywords Microblog analysis · Twitter · Agenda setting · Inter-policy agenda setting · Content analysis · Time series analysis

1 Introduction

The results of the United States presidential primaries in 2016 were highly unexpected. Donald Trump, a businessman with no prior experience in political office, was first seen as an outsider with no real chance at winning the nomination. "If Trump is nominated, then everything we think we know about presidential nominations is wrong", researchers at the University of Virginia Center for Politics said on their blog in August 2015 [28]. Trump was nominated and ultimately elected president, so what do we know?

J. Fromm (✉) · S. Melzer · B. Ross · S. Stieglitz
University of Duisburg-Essen, Duisburg, Germany
e-mail: jennifer.fromm@uni-due.de; bjoern.ross@uni-due.de; stefan.stieglitz@uni-due.de

© Springer International Publishing AG, part of Springer Nature 2018   201
R. Alhajj et al. (eds.), *Network Intelligence Meets User Centered Social Media Networks*, Lecture Notes in Social Networks,
https://doi.org/10.1007/978-3-319-90312-5_14

The 2016 primaries were exceptionally polarizing and emotional. Donald Trump insulted the other Republican candidates, the Democratic candidates, politicians from all over the world, the media, women and ethnic groups [21]. Researchers have argued that Trump offers the masculine image of a "tough guy" [32] and anti-politician who channels dissatisfaction [5]. It has also been argued in the media that Trump has been successful at using social media to rally his supporters [30]. However, conclusive scientific evidence for this assertion is so far missing.

It is clear, however, that both Trump and Clinton, or their respective campaign staff, are prolific and influential Twitter users. Clinton has more than 8 million followers and has written about 7000 short messages, or tweets, since 2013. Donald Trump has more than 10 million followers and has written more than 32,000 tweets since the creation of his account in 2009 (as of August 6, 2016). Both candidates are evidently able to reach millions of people on Twitter. Their tweets reflect their standpoints on political issues and their style of campaigning. In an analysis by the Washington Post of more than 6000 tweets posted by Donald Trump between June and December 2015, 11 % were found to be insulting [30].

The last decade has seen the emergence of social media and its acceptance as a new useful tool for researchers to examine a wide array of phenomena in domains such as politics and business [6, 12, 34]. In particular, it has helped researchers understand the dynamics of electoral campaigns [13, 20, 29, 40]. It has been studied how social media data can be used to predict voter behavior [23], how new technologies have shaped political campaigning [11], and which role social media appearances play in campaigns [4]. We contribute to this body of research by examining which issues were discussed on Twitter by the nominees in the run-up to the 2016 election, and how political issues were discussed.

To do so, we draw on agenda setting theory. This theory distinguishes three different agendas: the public agenda, the media agenda, and the policy agenda. For example, in this framework, the communication by the candidates on Twitter could be viewed as a reflection of the future policy agenda. Moreover, since communication on Twitter is generally public, candidates could influence each other in their communication and respond to one another. Therefore, this research addresses the following questions:

1. What is the nature of the Twitter communication by the main presidential candidates during the 2016 US primaries?
2. To what extent does agenda setting take place on Twitter between these candidates?

These research questions are addressed using a combination of content analysis and time series analysis. Content analysis is used to identify the most prominent topics discussed during the primaries by both eventual nominees (i.e., the first research question). Time series analysis is then used to address the second research question by examining the interrelations between the topic mentions over time. Prior research has investigated agenda setting on Twitter [10, 17], but the application of this combination of methods and a social media data source to study inter-candidate agenda setting is new.

In the remainder of the article, we give a theoretical overview and introduce the present literature on agenda setting. We especially focus on the policy agenda and Twitter. Our hypotheses are derived from the literature. The third section describes the methods used, and section four presents the empirical results. Section five contains a discussion of the results and our conclusions. We finally consider the limitations of this study and its implications for future research.

2 Related Work

Agenda setting theory concerns the idea that people do not only learn about a certain topic by media consumption, but also learn about its importance by evaluating the place and space of this specific topic [2]. In the words of Cohen [8, p. 13], media don't tell people "what to think", but "[. . .] what to think about". According to agenda setting theory, there are three different agendas: the public agenda, the media agenda, and the policy agenda. These agendas are interrelated and influence each other. Personal experience, interpersonal communication, and the "real world" have been found to influence all three agendas [14]. In the context of this study, the policy agenda is the most relevant.

2.1 The Policy Agenda

The policy agenda describes the actions taken by the government. The agendas of political parties, the bureaucracy, the President, the Committees, and the Lower and Upper House belong to the policy agenda [31]. In the policy agenda, issues are of great importance. It is crucial for the candidates to find the majority-efficient position for a certain issue to convince voters [19]. Candidates will give more salience to issues for which they get broader support from voters [9]. Studies about policy agenda setting pay attention to the dynamics in the political system and answer the question of how a new idea, policy, or problem is accepted in the political system [3].

The relationship between the media agenda and the policy agenda is reciprocal, that is, both agendas influence each other: Policy makers are not independent of the media—and the media are rarely independent of members of political institutions [41]. On the one hand, the president has been shown to influence the media agenda on foreign policy issues. In particular, issues with lower salience concerning foreign policy are most likely to be taken up by the media [25]. Moreover, the president has been found to be cited regularly during news about a press conference [15]. On the other hand, the media also influence the policy agenda by giving more attention to certain issues than to other issues. Media attention is often seen as "an agent of change" [41, p. 110] that has a stabilizing power for the policy-making process.

Politicians can take up issues from others and voice their opinions about these issues to distinguish themselves from other politicians. Soroka [31] calls these dynamics within political institutions *inter-policy agenda setting*. Studies regarding these dynamics found such effects between the President of the United States and Congress, with the former setting the agenda of the latter in most cases. Only for the issue of international affairs, Congress sets the President's agenda [27]. Moreover, the agenda setting process between candidates during election times has gained the attention of researchers during the last years. Inter-candidate agenda setting takes place for both issue salience and attribute salience relationships, but the results are stronger for attribute salience. The authors defined salience "by the frequency of issue and attribute mentions within campaign messages" [18].

2.2 Agenda Setting and Twitter

Campaign messages today are not only disseminated through traditional media, but also through social media such as the microblogging service Twitter. One can use the service to publish a status (*tweet*) with a maximum length of 140 characters. Users can also *follow* each other, that is, be notified when someone else publishes a tweet. Since its launch in 2006, Twitter has become a very popular medium in the USA, with 66 million monthly active users in June 2016 [33].

Even in times of a more fragmented media landscape, agenda setting takes place [36], but in a different way: Traditional media have become less powerful in the agenda-setting process, as their former power is now divided between traditional media and citizen media [24]. Since the election of Obama in 2008, Twitter has been widely used by politicians, especially during election periods [38]. Obama used Facebook and Twitter to collect donations and to connect to the community [26]. Politicians are known to have an influential role on Twitter in terms of retweets [13].

Research regarding agenda setting and Twitter has particularly focused on possible inter-media agenda setting effects. In a political context, it has been found that traditional media such as newspapers or television have a "symbiotic relationship" [10, p. 374] with the Twitter feeds of candidates and political parties for certain issues such as employment and health care during election times [10, 16].

Another research area concerning this topic is network agenda setting. This theory examines agenda setting effects between the media and public agenda from a network perspective. Using Big Data analysis, this theory has been confirmed for the 2012 election, when Mitt Romney was the Republican candidate who competed with Barack Obama. The authors found that the network issue agendas of the candidates' supporters correlated positively with the network issue agendas of certain media channels, in particular of horizontal media such as talk shows and cable news [37]. Especially young Americans who are part of the Twitter network also search for information on political issues and express their opinions [1].

In summary, Twitter is a useful and promising tool for studying agenda setting, but inter-policy agenda setting has not yet been explored in this context. We therefore test the following hypotheses on Twitter data.

2.3 Hypotheses

Inter-policy agenda setting can take place between parties. Vliegenthart et al. [39] examined the dynamics of the policy agenda in Belgium in a long-term study. They found that the parties influence each other and are more likely to take up the issues of other parties in parliament if they are from the same-language community. Moreover, governing parties have more agenda setting power than other parties. Hillary Clinton belongs to the governing party in the USA, was part of it as United States Secretary, and is also supported by the President. We therefore hypothesize that she will lead the agenda of Donald Trump:

H1 The issues mentioned in Hillary Clinton's tweets will predict the issues mentioned in Donald Trump's tweets.

Tedesco [35] found inter-candidate agenda setting effects in the candidate and campaign press releases of the Democratic candidates during the primaries in 2004. The author assumes that the agenda of the Democratic candidate John Kerry was set by the shared agenda of the other Democratic candidates. His opponent Howard Dean gained financial support and large media attention early, but he did not lead the other candidates' agendas. According to the author, Dean was not able to take advantage of the media attention. Furthermore, Vliegenthart et al. [39] find that extreme-right parties also have an agenda-setting power.

In 2016, Donald Trump received an exceptional amount of media attention, which he might be able to take advantage of. He has more Twitter followers than Clinton and makes polarizing statements that other candidates may have no choice but to react to. Thus, we hypothesize:

H2 The issues mentioned in Donald Trump's tweets will predict the issues mentioned in Hillary Clinton's tweets.

In summary, there are good reasons to believe that the candidates should influence each other's agenda. Examining whether this is indeed the case helps understand how Twitter was used by the candidates before the election.

3 Methods

To address the research questions and test the hypotheses, we chose a quantitative research design. A large dataset of relevant tweets was collected. Afterwards we conducted a content analysis to identify the most relevant political topics during the US primaries 2016. In a time series analysis, we focused on these identified topics and examined if any agenda setting effects occurred.

3.1 Dataset

As the present analysis focuses on the two main presidential primary candidates, only tweets and retweets by Hillary Clinton and Donald Trump (@HillaryClinton and @realDonaldTrump) were collected via the GET statuses/user_timeline endpoint of the Twitter API. This means we queried the API for all tweets and retweets sent from Donald Trump's and Hillary Clinton's accounts. Contrary to other research on Twitter, we did not search for tweets containing hashtags or keywords related to the candidates, as our research focused on the communication between the candidates and not on the communication of other Twitter users about them. Every tweet posted between November 15, 2015 and June 4, 2016 was collected, because the primaries took place within this timespan. Hillary Clinton occasionally tweets in Spanish. These tweets were excluded from further analysis. The cleaned dataset contained 6099 tweets. Of these, 3056 were posted by Donald Trump and 3043 by Hillary Clinton, so we were able to analyze a similar amount of tweets by each candidate.

3.2 Content Analysis

To address the first research question—that is, to examine the communication on Twitter by both candidates during the campaign—we conducted a content analysis of the tweets and determined the most salient topics in each candidate's messages. As the present dataset is too large to analyze them manually, the content analyis was conducted in a computer-assisted way.

The computer-assisted qualitative coding program QDA Miner and its text mining component WordStat were used to analyze the most frequent topics of both candidates. WordStat offers a few ready-to-use dictionaries. After testing those dictionaries on a small sample dataset, none of those dictionaries were found to be suitable for this research context. As political topics differ across elections, we used WordStat to develop a suitable dictionary ourselves.

The literature search revealed that Conway et al. [10] also conducted a computer-assisted content analysis to analyze agenda setting during US primaries. They used a dictionary including 21 political topics. These categories were used as a starting point for our own dictionary. First of all, all tweets were entered into QDA Miner and word frequencies were analyzed in WordStat. The most frequent words were classified and put into the dictionary categories based on the work of Conway et al. [10]. As expected, topics during these primaries differed from previous primaries, so not all categories by Conway et al. [10] were used and some new categories were added. As several words have different meanings, we resolved uncertainties by using the Keyword-in-Context tool. This shows all tweets including the word in question, so it is easier to decide into which category a word belongs.

After adding a decent amount of words, we conducted a first content analysis based on our own dictionary. WordStat was configured to use Porter Stemming and a built-in English exclusion list. Porter Stemming removes common English prefixes and suffixes before the categorization process. The exclusion list contained English stop words which provide no further meaning to a text (e.g., and, or, the). The built-in list was manually extended with Twitter-related stop words such as RT (the abbreviation of retweet). During the categorization process, the dictionary recognizes the beginnings and endings of words by identifying space characters. If a tweet contained one or more words included in the dictionary, it was assigned to all matching categories. Results returned a list of leftover words which were not included in the dictionary yet. We classified the most frequent leftover words to make sure that our dictionary contains all words occurring in more than one percent of all tweets.

Our final dictionary includes the following 21 categories: *employment, environment, guns, health care, military and defense, terrorism, slogans, media, family, human rights, meetings, thank-you messages, campaign funding, parties and politicians, caucus, foreign politics, education, economics, justice, Trump family* and *minorities*.

To validate our dictionary, we calculated recall and precision for each category. Therefore, 60 random tweets were coded manually first. Afterwards, the same tweets were coded by WordStat, using the developed dictionary. We calculated recall and precision values for each category. Scores for both measures can range from 0 to 1.0, where 1.0 would be the ideal result. A recall of 1.0 means that all tweets belonging to a specific category were labelled as belonging to this category by the dictionary but says nothing about how many other tweets were labelled incorrectly as belonging to this category. A precision score of 1.0 means that every tweet labelled as belonging to a specific category indeed belongs to this category but says nothing about the number of tweets that also belong to this category but were labelled incorrectly [22]. Most categories reached high values in recall and precision, but there are a few exceptions, for example the category *justice* ($R = 0.50$, $P = 1$). In sum, the dictionary reached an average recall of 0.84 and an average precision of 0.97 over all categories.

3.3 Time Series Analysis

To address the second research question and study the interrelations between the agendas of the two candidates, we used time series analysis. The content analysis served as a preprocessing step for the time series analysis. The agenda setting hypotheses were tested on political topics which were discussed frequently by both candidates. To identify these topics, the results of the content analysis were used.

During content analysis, the political topics *family, employment, minorities* and *terrorism* were identified as the ones most frequently discussed by both candidates. Therefore the dataset for time series analysis contained only tweets which were classified into at least one of these categories. For every candidate and topic, a time series was created resulting in eight different time series. All of these contain date and number of tweets related to a specific topic as variables. As Twitter is a very fast-paced medium, the number of tweets related to a specific topic was calculated for every day. The following method for analyzing time series data was also adapted from Conway et al. [10], as their research took place in a similar context.

The statistic software SPSS was used to analyze relationships between Donald Trump's and Hillary Clinton's time series. At first, all time series were tested for auto-correlations, which are correlations within a time series. Next, all time series were examined for linear trends by using curve estimation. For every time series with significant linear trends, SPSS automatically calculated a de-trended version of the original time series as a new variable. Finally, cross-correlations for every topic were calculated. Every time Donald Trump's time series was entered as the first variable and Hillary Clinton's as the second variable. In case a linear trend was found in the previous step, the de-trended time series was entered instead.

4 Results

In the following section results from content analysis and time series analysis are presented. At the end, we summarize shortly which hypotheses are supported by our results.

4.1 Content Analysis

The topic distributions of Donald Trump's and Hillary Clinton's tweets are shown in Table 1. It stands out that both candidates tweeted mostly about their media appearances, parties, and other politicians. Topics regarding political issues, for example, *employment*, *healthcare* or *human rights* were discussed much less often. *Employment* was the most frequently discussed political issue by Donald Trump, but occurred only in 4% of his tweets. This means that all other political issues were covered even less frequently. Hillary Clinton's topic distribution looks quite similar, but there are some differences. Her most frequently discussed political issue *family* occurred in 16% of her tweets. So both candidates had a different favourite political issue, and Hillary Clinton tweeted more often about her favourite issue *family* than Donald Trump tweeted about *employment*. Whereas Donald Trump discussed all political issues rarely, Hillary Clinton has one clear main issue but tweeted about all other political issues as seldom as Trump. *Minorities* as her second political issue, for example, accounted only for 6% of her tweets.

The main goal of the content analysis was to identify political topics appropriate for the subsequent time series analysis. These topics should occur quite often in both candidates' tweets, so that enough data for valid results are available. We decided to choose the most frequently discussed political topics from Trump's topic

Table 1 Topic distribution of Donald Trump's and Hillary Clinton's tweets in descending order of frequency

	Trump			Clinton		
	Category	No. of cases	% of cases	Category	No. of cases	% of cases
1	Parties and politicians	1783	58.36	Parties and politicians	1442	47.39
2	Media	778	25.47	Family*	488	16.04
3	Slogans	569	18.63	Media	446	14.66
4	Thank-you messages	493	16.14	Slogans	432	14.20
5	Caucus	296	9.69	Meetings	195	6.41
6	Employment*	129	4.22	Minorities*	189	6.21
7	Family*	92	3.01	Employment*	177	5.82
8	Terrorism*	83	2.72	Caucus	138	4.53
9	Trump family	78	2.55	Human rights	129	4.24
10	Minorities*	56	1.83	Healthcare	119	3.91
11	Healthcare	21	0.69	Thank-you messages	90	2.96
12	Military and defense	20	0.65	Environment	87	2.86
13	Foreign politics	13	0.43	Guns	85	2.79
14	Human rights	11	0.36	Terrorism*	68	2.23
15	Justice	8	0.26	Education	52	1.71
16	Campaign funding	7	0.23	Justice	49	1.61
17	Education	7	0.23	Economics	29	0.92
18	Economics	4	0.13	Military and defense	19	0.62
19	Guns	4	0.13	Campaign funding	6	0.20
20	Environment	2	0.07	Foreign politics	5	0.16
21	Meetings	0	0.00	Trump family	0	0.00

Categories marked with * were the most frequently discussed political topics and used in further analysis

distribution and checked if they also occur often enough in Clinton's tweets. As a result, the political topics *employment, family, terrorism* and *minorities* were chosen for further analysis.

4.2 Time Series Analysis

As described in Sect. 3, every time series was tested for auto-correlations and linear trends. In the following time series, significant auto-correlations were found: Trump family (lag 1, lag 2), Trump terrorism (lag 1, 2, 3 and 14), Clinton minorities (lag 1),

Table 2 Cross-correlation results for the topics minorities, terrorism, employment and family

Political topic	Significant lags and leads for Trump with Clinton	Cross-correlation coefficient
Minorities	Lag (−4)	0.20
Terrorism	Lag (−3)	0.16
	Lag (−1)	0.32
	Lead (2)	0.19
	Lead (4)	0.18
Employment	No significant lags or leads	
Family	No significant lags or leads	

and Clinton family (lag 14). Significant linear trends were found in these time series: Trump minorities ($R^2 = 0.03$, $p < 0.05$), Trump terrorism ($R^2 = 0.05$, $p < 0.01$), Clinton family ($R^2 = 0.02$, $p = 0.05$), and Clinton terrorism ($R^2 = 0.03$, $p = 0.01$).

In the next step, cross-correlation coefficients were calculated for all four topics. Table 2 shows that for the topic *minorities*, only the cross-correlation coefficient for lag −4 was significant. This means that Trump's tweets about minorities were predicted by Clinton's tweets about this topic 4 days earlier. Table 2 also shows that for the topic *terrorism*, cross-correlation coefficients for several lags and leads were significant (lag −3 and −1, lead 2 and 4). These results imply that in contrast to the topic *minorities*, correlations between both candidates' time series are bi-directional. In other words, Trump's tweets about terrorism were predicted by Clinton's tweets about this topic 1 and 3 days earlier, but Clinton's tweets were also predicted by Trump's tweets 2 and 4 days earlier. For the other topics (*family* and *employment*), no significant cross-correlation coefficients were found.

In Figs. 1 and 2, visualizations of the *minorities* and *terrorism* time series are shown, as cross-correlation returned significant results for these topics. It stands out that both candidates tweet much less continuously about terrorism. Most of the time numbers of daily tweets are quite low, but four different peaks can be identified during the tracking timespan. Figure 2 shows that terror attacks took place right before the first three peaks. Since many significant auto-correlations were found in Trump's *terrorism* time series and terror attacks took place right before the peaks, the significant cross-correlation should be interpreted carefully. It is possible that these external events instead of agenda setting explain the correlations between Trump's and Clinton's time series for the topic *terrorism*. For the topic *minorities* no such external events were found, so agenda setting might be a reasonable explanation for the significant cross-correlation.

Table 3 summarizes the results of the hypothesis tests. The findings provide partial support for the first hypothesis, and no support for the second hypothesis.

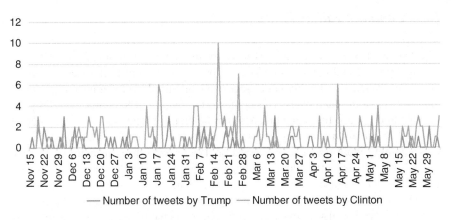

Fig. 1 Visualization of both candidates' time series for the topic *minorities*

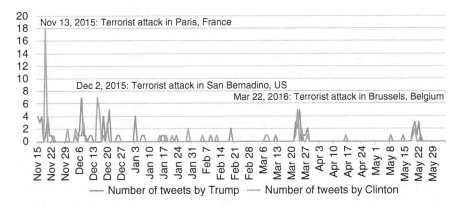

Fig. 2 Visualization of both candidates' time series for the topic *terrorism*

Table 3 Results of hypothesis tests

	Hypothesis	Result
H1	The issues mentioned in Hillary Clinton's tweets will predict the issues mentioned in Donald Trump's tweets	Supported for topic *minorities*
H2	The issues mentioned in Donald Trump's tweets will predict the issues mentioned in Hillary Clinton's tweets	Not supported

5 Discussion

This section discusses the results and addresses the research questions. It also addresses the limitations of the study and mentions implications for future research.

5.1 Communication Topics

The first goal of this research was to study the nature of communication by the eventual presidential nominees on Twitter. We found that political issues on Twitter are not as frequent as thank-you messages and announcements for upcoming media appearances of the candidates. This confirms the finding of Sandoval et al. [29].

This finding also has some practical implications. Twitter serves as a source of political information for many people [1], so politicians could likely benefit from disseminating more political statements on Twitter than they currently do. On the other hand, Twitter only allows messages with up to 140 characters, which may make it difficult to discuss complex topics. Secondly, during his campaign, Donald Trump was known for his plan of building a wall between Mexico and the USA to stop crime. He was also frequently mentioned in the media for his plans to institute a law that would make the immigration of Muslims into the USA illegal [32]. It can therefore be considered somewhat surprising that the number of tweets from Donald Trump about minorities was very low. There were only 56 tweets regarding this issue in the tracking period. An explanation for this divergence might be inter-media agenda setting. Perhaps, the statements were so arousing or polarizing that journalists picked up on them and wrote or talked about them much more than Donald Trump did himself.

A comparison of the candidates regarding specific topics reveals that both politicians took them up differently. For example, after the terrorist attack of San Bernadino, the biggest terror attack in the USA [7], Donald Trump criticized that it was not reported as a terrorist act (December 4, 2015). Hillary Clinton claimed that the attack could only take place because "Republican Senators blocked a bill to stop suspected terrorists from buying guns" (December 4, 2015). Here, Hillary Clinton uses the terror attack to criticize the Republican Party. Donald Trump also blamed Hillary Clinton indirectly in March, during the Brussels terror attacks: "Hillary Clinton has been working on solving the terrorism problem for years. TIME FOR CHANGE, I WILL SOLVE - AND FAST" (March 24, 2016). During these bombings, Clinton twittered: "These terrorists seek to undermine the democratic values that are the foundation of our way of life. They will never succeed. -H." (March 22, 2016). All in all, both candidates used terrorist attacks for their purposes. When comparing the issue *family*, it becomes clear that the perception that Hillary Clinton plays the "woman card" [42] may be rooted in the fact that she addresses this topic five times as often as Trump.

5.2 Agenda Setting

In addressing the second research question, this paper is the first one to examine inter-policy agenda setting effects on the microblogging service Twitter for candidates during US primary elections. The goal was to answer the question whether agenda setting takes place between the candidates during the US presidential elections. This question can only partially be answered with yes. For the topic *minorities*, Clinton is leading Trump's agenda. Results for *employment* and *family* were not significant. We conclude that the presence of agenda setting depends on the political issue. With this research, we contribute to the work of Soroka [31], who introduced the notion of inter-policy agenda setting and confirmed existing research by Vliegenthart et al. [39].

For the topic *terrorism*, agenda setting might have taken place in both directions, but significant auto-correlations were present even after removing linear trends, and the cross-correlations observed may therefore be spurious. Figure 2 shows that attacks took place right before the peaks, which may explain this phenomenon. We therefore refrain from drawing further conclusions from the statistical results.

A closer inspection of the tweets reveals that when Donald Trump takes up the issues of other candidates and politicians, he frequently does so to attack them. For example, Trump twittered on December 7, 2015: "Obama said in his speech that Muslims are our sports heroes. What sport is he talking about, and who? Is Obama profiling?" In this tweet, he takes up Obama's speech to ask questions, encouraging his community to think about it.

In summary, this study offers some evidence that agenda setting actually takes place on Twitter, but less than could have been expected. In some cases (e.g., *terrorism*), tweets looked more likely to be prompted by external events. This finding is important because it emphasizes further how little interaction with the political opponent occurred on Twitter in this case. Thereby our research contributes to the small existing body of literature on agenda setting in the context of Twitter.

5.3 Limitations

As any research, ours comes with limitations. Spanish tweets were not considered in this analysis. Expanding the dictionaries with Spanish words would have been very time-consuming. However, since Clinton sometimes tweeted in Spanish, this decision could skew the results on the topic of *minorities*.

Furthermore, when creating a dictionary for content analysis, there are always ambiguities regarding the appropriate category for a particular word, and building a dictionary is a process that involves subjective decisions. We made use of the Keyword-in-Context tool to resolve these uncertainties and evaluated the dictionary by calculating recall and precision for every category. While validation results should be interpreted carefully, they suggest that the dictionary was suitable for our purposes.

We found significant auto-correlations in some time series, which led us to exclude the topic of *terrorism*. Linear trends were controlled for by using de-trended versions of the affected time series as described in Conway et al. [10], but more sophisticated methods are available to remove auto-correlations.

Finally, as already mentioned, the results of the US primaries in 2016 were highly unexpected. Almost no one thought that Donald Trump will be the candidate for the Republican Party. This provided a unique research setting, but it is uncertain if this research is replicable for other contexts and for other countries.

5.4   Future Research

We propose further research to examine if the results can be generalized. Additionally, given the highly emotional content of tweets, it seems worthwhile to examine if emotion has an influence on agenda setting on social media, especially on Twitter. For example, the difference between candidates regarding the sentiment of their tweets could be examined in future research. It is also unclear whether sentiment has an influence on agenda setting on Twitter.

While executing our time series analysis, we found auto-correlations in Trump's series for *terrorism*. Possible external events are predicting the time series of Trump. In further research, a more complex model could be developed that removes these auto-correlations or explicitly takes the influence of outside events into account. Another valuable research direction would be the enhancement of automatic classification methods which allow the identification of political topics in election-related tweets even though topics change between elections.

Future research should also include tweets by the public to evaluate how agenda setting takes place between politicians and citizens on twitter. However, in this regard it has to be considered that Twitter is only used by a small percentage of the whole population and mostly by younger people. Once again, the influence of sentiment should be considered in this context.

6   Conclusions

In his 1996 analysis of how a new technology had reshaped political campaigning in Texas, Jonathan Coopersmith stated that "the spread of modern information technologies has greatly altered the face of politics" [11, p. 37]. The technology he examined was the fax machine. Twenty years later, many of the observations he made are equally true for Twitter: a flood of data is being generated, information can be disseminated rapidly, and there is consequently increased pressure on political campaigns to make use of these new technologies effectively.

But Twitter does not only enable campaigns to spread information rapidly, it also allows researchers an unprecedented glimpse at the daily activities of campaigns. When Wattal et al. [40] laid out their research agenda, they called on researchers to examine how the political system might change as a result of the Internet. They ask, "how might the web be used to support increased mutual understanding and tolerance in political discourse"?

Twitter has indeed become one of the most important social media used by campaigns—but our analysis showed that very little political discourse actually takes place there. The medium is dominated by thank-you messages and simple political slogans. Candidates use it to reach out to their followers, not to engage with the political opposition.

Still, the few political messages present in the data open new avenues for researchers. Previous research on agenda setting had considered the agenda of political actors and institutions in power. Now, a large part of the communication by political campaigns is readily available to researchers in a digitized form. We can analyze the agenda of those who will wield political power even before they do it. Twitter has made it possible to carry out this analysis with less effort and at a larger scale than before.

In this study, we combined content analysis and time series analysis. Through the resulting analysis of day-to-day frequencies of topic mentions, we were able to find little evidence for inter-policy agenda setting during the run-up to the US presidential elections. Instead of fostering discussion and helping mutual understanding, Twitter seems to represent a fractured social space.

References

1. Ancu, M., Cozma, R.: MySpace politics: uses and gratifications of befriending candidates. J. Broadcast. Electron. Media **53**(4), 567–583 (2009)
2. Baran, S.J., Davis, D.K.: Mass Communication Theory: Foundations, Ferment, and Future, vol. 6. Wadsworth, Boston (2015)
3. Baumgartner, F.R., Green-Pedersen, C., Jones, B.D.: Comparative studies of policy agendas. J. Eur. Public Policy **13**(7), 959–974 (2006)
4. Bühler, J., Bick, M.: The impact of social media appearances during election campaigns. In: AMCIS Proceedings, vol. 5, pp. 3406–3416 (2013)
5. Button, M.E.: Trump and the triumph of hubris over democratic politics (2016). https://global. oup.com/academic/category/social-sciences/politics/2016-election/tattohodp/
6. Cazzoli, L., Sharma, R., Treccani, M., Lillo, F.: A large scale study to understand the relation between twitter and financial market. In: ENIC Proceedings, pp. 98–105 (2016)
7. CNN: San Bernadino Shooting (2016). http://edition.cnn.com/specials/san-bernardino-shooting
8. Cohen, B.C.: The Press and Foreign Policy. Princeton University Press, Princeton (1963)
9. Colomer, J.M., Llavador, H.: An agenda-setting model of electoral competition. SERIEs **3**(1–2), 73–93 (2012)
10. Conway, B.A., Kenski, K., Wang, D.: The rise of twitter in the political campaign: searching for intermedia agenda-setting effects in the presidential primary. J. Comput. Mediat. Commun. **20**(4), 363–380 (2015)
11. Coopersmith, J.: Texas politics and the fax revolution. Inf. Syst. Res. **7**(1), 37–51 (1996)
12. Cossu, J.V., Dugue, N., Labatut, V.: Detecting real-world influence through twitter. In: ENIC Proceedings, pp. 83–90 (2015)
13. Dang-Xuan, L., Stieglitz, S., Wladarsch, J., Neuberger, C.: An investigation of influentials and the role of sentiment in political communication on twitter during election periods. Inf. Commun. Soc. **16**(5), 795–825 (2013)

14. Dearing, J.W., Rogers, E.M.: Communication Concepts 6. Agenda-Setting. Sage, Thousand Oaks (1996)
15. Eshbaugh-Soha, M.: Presidential influence of the news media: the case of the press conference. Polit. Commun. **30**(4), 548–564 (2013)
16. Groshek, J., Clough Groshek, M.: Agenda trending: reciprocity and the predictive capacity of social networking sites in intermedia agenda setting across topics over time. Media Commun. **1**(1), 15–27 (2013)
17. Gruszczynski, M.W.: Examining the role of affective language in predicting the agenda-setting effect. In: APSA Annual Meeting (2011)
18. Kiousis, S., Shields, A.: Intercandidate agenda-setting in presidential elections: issue and attribute agendas in the 2004 campaign. Public Relat. Rev. **34**(4), 325–330 (2008)
19. Krasa, S., Polborn, M.: The binary policy model. J Econ. Theory **145**(2), 661–688 (2010)
20. Larsson, A.O., Moe, H.: Studying political microblogging: Twitter users in the 2010 Swedish election campaign. New Media Soc. **14**(5), 729–747 (2012)
21. Lee, J.C., Quealy, K.: The 337 people, places and things Donald trump has insulted on twitter: a complete list (2016). https://www.nytimes.com/interactive/2016/01/28/upshot/donald-trump-twitter-insults.html
22. Liu, B., Hu, M., Cheng, J.: Opinion observer: analyzing and comparing opinions on the web. In: WWW Proceedings, pp. 342–351 (2005)
23. Maldonado, M., Sierra, V.: Can social media predict voter intention in elections? the case of the 2012 dominican republic presidential election. In: AMCIS Proceedings (2015)
24. Meraz, S.: Is there an elite hold? traditional media to social media agenda setting influence in blog networks. J. Comput. Mediat. Commun. **14**(3), 682–707 (2009)
25. Peake, J.S.: Presidential agenda setting in foreign policy. Polit. Res. Q. **54**(1), 69–86 (2001)
26. Robertson, S.P., Vatrapu, R.K., Medina, R.: Off the wall political discourse: facebook use in the 2008 U.S. presidential election. Inf. Polity **15**(1–2), 11–31 (2010)
27. Rutledge, P.E., Larsen Price, H.A.: The president as agenda setter-in-chief: the dynamics of congressional and presidential agenda setting. Policy Stud. J. **42**(3), 443–464 (2014)
28. Sabato, L.J., Kondik, K., Skelley, G.: Republicans 2016: what to do with the donald? (2015). http://www.centerforpolitics.org/crystalball/articles/republicans-2016-what-to-do-with-the-donald/
29. Sandoval, R., Matus, R.T., Rogel, R.N.: Twitter in Mexican politics: messages to people or candidates? In: AMCIS Proceedings, pp. 1–10 (2012)
30. Schwartzman, P., Johnson, J.: It's not chaos. It's Trump's campaign strategy (2015). https://www.washingtonpost.com/politics/its-not-chaos-its-trumps-campaign-strategy/2015/12/09/9005a5be-9d68-11e5-8728-1af6af208198_story.html
31. Soroka, S.N.: Issue attributes and agenda-setting by media, the public, and policymakers in Canada. Int. J. Public Opin. Res. **14**(3), 264–285 (2002)
32. Sperling, V.: Masculinity, misogyny, and presidential image-making in the U.S. and Russia (2016). https://global.oup.com/academic/category/social-sciences/politics/2016-election/mmpimur/
33. Statista: Number of monthly active twitter users in the United States from 1st quarter 2010 to 2nd quarter 2016 (in millions) (2016). https://www.statista.com/statistics/274564/monthly-active-twitter-users-in-the-united-states/
34. Stieglitz, S., Dang-Xuan, L., Bruns, A., Neuberger, C.: Social media analytics—an interdisciplinary approach and its implications for information systems. Bus. Inf. Syst. Eng. **56**(2), 101–109 (2014)
35. Tedesco, J.C.: Intercandidate agenda setting in the 2004 democratic presidential primary. Am. Behav. Sci. **49**(1), 92–113 (2005)
36. Vargo, C.J.: Twitter as public salience: an agenda-setting analysis. In: AEJMC Annual Conference (2011)
37. Vargo, C.J., Guo, L., Mccombs, M., Shaw, D.L.: Network issue agendas on twitter during the 2012 U.S. Presidential election. J. Commun. **64**(2), 296–316 (2014)
38. Vergeer, M.: Twitter and political campaigning. Sociol. Compass **9**(9), 745–760 (2015)

39. Vliegenthart, R., Walgrave, S., Meppelink, C.: Inter-party agenda-setting in the Belgian parliament: the role of party characteristics and competition. Polit. Stud. **59**(2), 368–388 (2011)
40. Wattal, S., Schuff, D., Mandviwalla, M., Williams, C.B.: Web 2.0 and Politics: the 2008 US presidential election and an e-politics research agenda. MIS Q. **34**(4), 669–688 (2010)
41. Wolfe, M.: Putting on the brakes or pressing on the gas? Media attention and the speed of policymaking. Policy Stud. J. **40**(1), 109–126 (2012)
42. Zezima, K.: Trump: Clinton is playing the 'woman card' (2016). https://www.washingtonpost.com/news/post-politics/wp/2016/04/26/trump-clinton-is-playing-the-woman-card/

Strength of Nations: A Case Study on Estimating the Influence of Leading Countries Using Social Media Analysis

Alexandru Topîrceanu and Mihai Udrescu

Abstract The use of social media has become pervasive for social networking and content sharing. Every day, thousands of news articles are written about events occurring around the world. And yet, the content that is generated from these websites is so massive that a large proportion of the data remains largely neglected. Similar to how users augment their influence by sharing content, online published mass-media can model the flow of influence between countries. Our work bridges the gap between political and economic analysis, by relying on social media to understand key aspects of relationships between countries in various contexts of interest. We use five datasets of influential countries, combined with seven hot topics in news, and use Event Registry as a data mining tool; out of the gathered data we further build directed weighted graphs. Based on the topological structures, we introduce four different node strength measures and showcase different snapshots of influence between countries.

Keywords Social media analysis · Influence · European Union · GDP · MOOC · Graph metric

1 Introduction

Over the past decade, social media has erupted as a category of online discourse where individuals create and share content over the network, at a remarkable rate. Examples include Facebook, Twitter, MySpace, Snapchat, LinkedIn, etc.

There are thousands of news articles written and published every day by news agencies all across the world, and this information dynamics can be analysed to better understand the impact that an event from one country has on the media of

A. Topîrceanu (✉) · M. Udrescu
Department of Computer and Information Technology, "Politehnica" University Timisoara, Timisoara, Romania
e-mail: alext@cs.upt.ro; mudrescu@cs.upt.ro

© Springer International Publishing AG, part of Springer Nature 2018
R. Alhajj et al. (eds.), *Network Intelligence Meets User Centered Social Media Networks*, Lecture Notes in Social Networks,
https://doi.org/10.1007/978-3-319-90312-5_15

another. From a higher level standpoint, it can be interpreted that if a news makes international headlines in other countries, then we consider that the event-source country has an influence over all other countries which relate to that particular event in their media. Quantifying the amount of news that makes it through to another country, we can also estimate the weight of the impact.

In order to acquire and process all relevant social media data we turned to the Event Registry platform, developed by a team of researchers at Jozef Stefan Institute in Slovenia [8]. We process the gathered data into weighted directed graphs for several groups of countries of interest, and focus on a predefined set of discussion topics, or tags. Through this analysis, we are able to show how insightful the processing of mass media can be, using a case study for the influence of EU countries in terms of GDP and economics. We also compare the graph metric observations with datasets like the G8 and G20 countries, and discuss how the world economic scene is more decentralized than continental environments, such as the EU.

Finally, the paper introduces four possible measures for quantifying the strength of a nation, and discusses the advantages for each of them in turn. We show that considering weighted in- and out-degree offers the most intuitive results in terms of estimating how much influence a country dissipates over other countries. The methodology discussed here may benefit further disciplines such as political studies, open education, MOOCs, and communication studies, which often work with more qualitative approaches.

2 State of the Art

We rely extensively on centrality distributions in order to analyse the modelled networks in terms of structural properties and to estimate the influence of nodes. The most important centrality measures we use to estimate influence are: (weighted in/out) degree, betweenness, PageRank, HITS and eigenvector centrality [10, 20].

There are noteworthy studies in the direction of topological network analysis which explain the function of the underlying modelled graph. To mention a few, there are studies empowering the reproduction of reliable recommender systems [11], the so-called recipe network [13], and even a network for characters in the Marvel universe [1]; all these analyse the human desire of sharing information based on particular interests. Other studies include an overview of the professional collaborations within the global music industry, in the so-called MuSeNet [18], and an original perspective on the female fashion world [14], with FMNet built as a physical similarity network which explains how trends and icons are formed in fashion.

Last, in the field in which we apply network analysis, there are some interesting results, but they originate from socio-political backgrounds, using different methods and tools than the ones we employ here [5–7]. It was however shown that, in some particular cases, social media can be successfully used to understand, model and even predict user behaviour [6, 19].

3 Methods

3.1 Event Registry

Event Registry is a platform developed by Leban et al. which can automatically analyse news articles, in order to identify mentioned world events. Their developed system is able to pinpoint groups of articles that describe the same event, by aggregating articles even from different languages. The collected articles originate from about 75,000 news sources, and their number ranges between 100,000 and 200,000 articles per day. English, German, Spanish and Chinese are the only languages that are syntactically and semantically processed. Events are constructed by finding groups of articles describing the same event, extracted from the articles event information. In order to identify groups of articles describing the same event, an online clustering algorithm is used. Furthermore, we needed metadata information from the news data, so we used the API to extract the events core information, such as event location, date, who is involved, and what is it about, from every article.

Event Registry provides four different pricing plans, from free to custom, with a limit on the amount of searches that can be queried (per month). For this study, we used the free price plan which includes 2000 searches.

3.2 Data Acquisition

We access data from Event Registry through the available API by issuing HTTP GET requests with specific parameters. To this end, we used the provided Python SDK, to which contributions were eventually also made via GitHub. The SDK provides classes and methods that automatically generate the necessary requests in order to obtain the data. When accessing data in Event Registry, one needs to specify an API key so that the number of requests made by a user is known.

Our study focuses on understanding how various countries in Europe, and around the world, are bridged through social media and mass-media news. Scanning for all news mentions between all countries around the world would be limited by two factors: the amount of queries that we can send to Event Registry, and the limited amount of news that is available from some countries. Trying to cover all countries would most likely result in a disconnected graph, hence this study relies on the following five sets of countries: EU (28 member states), G8 (eight most powerful countries and EU), G8+5 (G8 plus 5 emerging countries), G20 (top 20 major economies and central bank) and NATO.

Second, we define several news tags for our queries. A tag is a keyword that can generically describe a news item, or a world event extracted by Event Registry. The tags can be specified in the query string, and the online platform automatically filters the necessary information. In order to obtain a multilateral perspective on the influence and strengths of the nations listed above, we extract social media information based on the following seven tags: Economics, Finance, GDP, Industry, Military, Politics, Warfare.

As such, we obtain 5×7 independent raw datasets, by combining all country sets with all tags (e.g. EU_finance, EU_gdp, G8_gdp, etc.).

3.3 Graph Modelling and Representation

In order to build a graph out of the data extracted from Event Registry, we implement the following intuitive approach. Each obtained news item has a tag (e.g. sports) and a location in which it happened (e.g. "Canadian Formula 1 GP on June 12th won by Lewis Hamilton" is located in Canada). Each single news item can be mentioned by other countries. For example, if a French media agency mentions the Canadian Formula 1 GP, then we interpret this as "*Canada* has an influence over *France*, in the field of *sports*". Similar, any other country writing about the same news will have been influenced by the news source. Furthermore, a country may write multiple articles about the same event (e.g. a stock market decrease), so that the relationship becomes weighted.

Formally, we define a graph $G = (V, E)$, consisting of countries G, and the weighted directed edges E between them. A directed edge $e_{ij} \in E$ is added between a source node n_i and target node n_j if there is specific news that occurred in n_j about which n_i wrote. The number of such news, or the number of mentions of a single piece of news represents the weight w_{ij} of edge e_{ij}. Additionally, there might exist a different directed weighted edge e_{ji}, or none at all. The methodology for building G is described in Fig. 1.

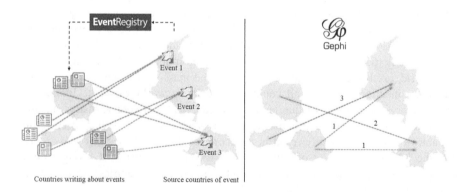

Fig. 1 Construction of an influence graph between countries based on the social media exchange between them (**left**). Countries become nodes and mentions become directed edges (**right**). The direction of an edge in graph G is always pointing from a country mentioning an article, towards the news source country; the weight is proportional to the amount of mentions directed towards the same event source country

The obtained data is converted in *gdf* file format and processed with Gephi [2]. For placement and visualization we use the existing ForceAtlas2 algorithm [3], and modularity [9] for graph partitioning into communities [4].

3.4 Defining the Strength of a Nation

There is a variety of analytical metrics inspired by topological analysis, such as the Erdos number [10] (based on shortest path length), the fractal dimension d_B [12] (based on the box counting method to estimate the structural complexity order of a graph), the sociability metric S [18] (based on graph metrics applied on an ego-network of friends), or the fidelity metric φ_A [15, 16] (based on common topological features of complex networks), we make the following considerations in our definition of σ as the strength of a nation.

To this end, we introduce a set of intuitive metrics that quantify the influence of nations. We propose four possible definitions for σ, based on gauging different node centrality measures; then we present an exploratory analysis in order to reveal which one corresponds to the most useful and insightful definition. On the one hand, we consider that higher centralities of a node in a network mean a higher overall (or average) influence of that node. As such, we define σ_{all} as:

$$\sigma_{all} = \begin{cases} WID/WOD \times Btw \times PR \times HA \times Eg, & \text{if } WOD > 0 \\ 0, & \text{if } WOD = 0 \end{cases} \quad (1)$$

In Eq. (1) the involved graph metrics are: weighted in-degree (WID) and out degree (WOD), normalized betweenness (Btw), PageRank (PR), HITS authority (HA) and Eigenvector centrality (Eg). While this definition of σ produces the most composite overview of a country node in a network, it has the disadvantage that any centrality equal to 0 will render the result 0, so many nodes will be rendered with $\sigma_{all} = 0$.

Our next consideration is that in a graph of this type, all centralities except degree (in/out) are less relevant, so we define the strength based on weighted degree solely. The idea is that node betweenness, for example, does not imply any control over communication in this context, since all countries may be connected to any other country through a direct link, if news was shared. In other words, each graph has the possibility of being fully connected, so a high PageRank or betweenness for a node does not suggest increased strength. As such, we define the following three strengths:

$$\sigma_{i/o} = \begin{cases} WID/WOD, & \text{if } WOD > 0 \\ 0, & \text{if } WOD = 0 \end{cases} \quad (2)$$

$$\sigma_{i+o} = WID + WOD \quad (3)$$

$$\sigma_{comp\_io} = \begin{cases} WID/WOD \times (WID + WOD), & \text{if } WOD > 0 \\ 0, & \text{if } WOD = 0 \end{cases} \tag{4}$$

The rationale behind Eq. (2) is that a higher in-degree means higher influence, conditioned by a lower out-degree. On the other hand, often "more is better", in terms of news mediation also; as such, Eq. (3) assigns higher strength to countries with overall high degree. Indeed, Eq. (4) creates a sweet-spot between the previous two definitions, considering both a high in-degree and consistent quantitative mediation matter.

In other words, we use the notation $\sigma_{set}^{tag}(nation)$, e.g. $\sigma_{EU}^{politics} Belgium$, to express the strength of a *nation* inside a given group of countries (e.g. EU, NATO), and for a specific tag only (e.g. politics).

4   Results

4.1   Individual Metric Analysis Case Study

We introduce 10 distinct datasets, and their representative graph measurements in Table 1. While none of the graphs can be described as "large and complex networks", they do possess representative distributions that are naturally found in complex social networks [17]. The average path length and clustering coefficients are abiding by both small-world and scale-free principles [10], namely high clustering and low path length are present in all our modelled networks.

We first notice a higher AD in G8 and G20 datasets, as these nations share media between them forming an almost complete graph. We corroborate this observation

Table 1 Graph measurements for 10 relevant datasets identified by name of the country group (e.g. EU, NATO) and search tag (e.g. economics), average degree (AD), density (Dns), clustering coefficient (ACC), and average path length (APL)

Dataset	Nodes	Edges	AD	Dns	ACC	APL
EU_economics	21	53	2.76	0.126	0.291	2.031
EU_gdp	28	132	7.32	0.175	0.517	1.839
EU_industry	28	236	18.43	0.312	0.581	1.717
EU_warfare	28	109	6.68	0.144	0.593	2.014
NATO_economics	19	47	3.21	0.137	0.479	1.904
NATO_gdp	27	166	11.15	0.236	0.572	1.178
NATO_industry	28	284	45.39	0.376	0.607	1.597
NATO_warfare	27	131	10.18	0.187	0.65	1.923
G8_industry	9	63	104.9	0.875	0.875	1.016
G20_industry	20	297	121.8	0.782	0.823	1.177

with the fact that, while trade is decentralized at a global scale, smaller economical-systems, like the EU, are still highly centralized due to the presence of a few dominant countries, which makes us conclude that inter-continental economics is more meritocratic in its structure, while intra-continental economical structures and mechanisms are more topocratic.

Further, we analyse the distribution of node centralities in the *EU* dataset, by taking into consideration the "gdp" and "economics" tags. The most relevant observations extracted from social media are depicted in Fig. 2a–d. All nodes are sized and coloured proportionally to the centrality highlighted in each panel, while their placement is roughly geographical; all centralities are measured using the plug-ins available in Gephi [2]. By examining each sub-figure, we can notice different types of influence within the EU.

UK's dominant betweenness stems from the tight interconnection (i.e. high edge weights) with some countries from different geographical regions of Europe: Spain, Finland, Italy, Belgium and Cyprus. This means that media sharing often uses UK as a hub. This in turn is explained by the British origin of some of the most credited media agencies in Europe, like Reuters and BBC. In terms of both PageRank and HITS, France, alongside Germany and the UK, clearly stand out. The trio emerges as a media clique, as there are high edge weights between all of them, which increases the relative influence of all of them. We believe France is ranked first by these centralities because of its overall higher degree. The news agency France-Press consistently covers most news around Europe (e.g. Euronews). In terms of weighted degree, there are two different perspectives presented in Fig. 2c, d. France has the highest out-degree, and the explanation is correlated with the previous observation about PageRank and the activity of France-Press. On the other hand, Belgium has the highest in-degree, which is likely determined by Bruxelles' position as EU capital. Most countries—especially from the Eastern block—link to Belgium by mentioning both the keywords *EU* and *GDP*.

From these results we emphasize that social media can, indeed, offer insightful explanations for the relationships between states. However, for the best socio-political interpretation, one needs to study multiple node centralities and have the appropriate background knowledge.

4.2 Measuring Nation Strength

Next, we present the measurements for nation strength, according to Eqs. (1)–(4), and discuss which is the most representative and insightful of the formulae. In Table 2, we present correlations between each σ strength and other centralities. In Table 3, we present three sets of measurements for $\sigma_{EU}^{economics}$, $\sigma_{EU}^{warfare}$ and $\sigma_{G20}^{industry}$. It comes as no surprise that we obtain fairly different rankings for each definition of strength.

EU_gdp

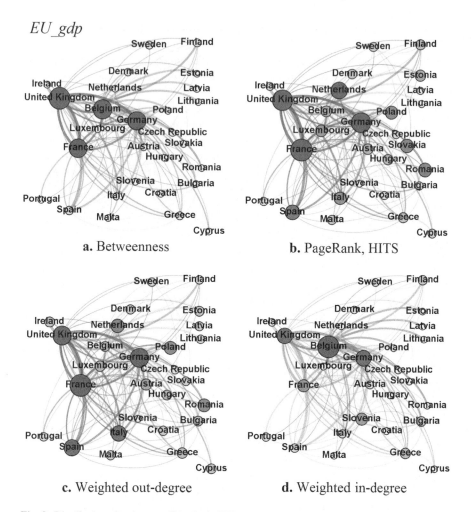

Fig. 2 Distribution of node centralities in the EU countries, where nodes are placed geographically and are sized proportional to each used centrality, and edges represent directed mentions in mass-media containing the tag "GDP"

The most stable results are obtained for the *EU_economics* dataset, where UK and Germany constantly rank in the top. Analysing the top panel in Table 3, we consider that $\sigma_{i/o}$ yields slightly different results than the other three definitions, and that σ_{all} produces a distribution that decreases too abruptly (see that UK ranks first with $\sigma_{i/o} = 1$, followed by 0.463 and 0.052). As such, we consider the distribution of σ_{all} too abrupt to make it an intuitive tool for analysis. In the middle panel, we notice that σ_{i+o} produces a slightly different ranking than the others. Also, σ_{all} is again more abrupt than the other distributions, and $\sigma_{i/o}$ introduces rather unexpected countries in the ranking (i.e., in terms of *warfare*). We consider σ_{i+o} and σ_{comp_io} to be more insightful in this second case.

Table 2 Correlations between σ metrics and other graph centralities (in/out-degree, betweenness, PageRank) on the *EU_economics* dataset

	σ_{all}	$\sigma_{i/o}$	σ_{i+o}	σ_{comp_io}	WID	WOD	Btw	PR
σ_{all}	–	0.704	0.455	0.802	0.730	0.442	0.815	0.647
$\sigma_{i/o}$		–	0.450	0.747	0.625	0.878	0.797	0.524
σ_{i+o}			–	0.788	0.456	0.288	0.313	0.301
σ_{comp_io}			–	–	0.755	0.480	0.603	0.558

Table 3 Ranking of top five nations in terms of each σ strength for three datasets: A (*EU_economics*), B (*EU_warfare*), C (*G20_industry*)

A	σ_{all}	$\sigma_{i/o}$	σ_{i+o}	σ_{comp_io}
1	**UK** (1.000)	Czech R. (1.000)	**UK** (1.000)	**Germany** (1.000)
2	**Germany** (0.463)	**Germany** (0.800)	**Germany** (0.765)	UK (0.726)
3	Poland (0.052)	**UK** (0.444)	France (0.647)	France (0.302)
4	France (0.003)	France (0.286)	Poland (0.647)	Czech R. (0.288)
5	Hungary (0.002)	Hungary (0.250)	Ireland (0.647)	Poland (0.198)
B	σ_{all}	$\sigma_{i/o}$	σ_{i+o}	σ_{comp_io}
1	**Netherl.** (1.000)	**Netherl.** (1.000)	**Germany** (1.000)	**Netherl.** (1.000)
2	UK (0.570)	Sweden (0.857)	UK (0.653)	Sweden (0.375)
3	**Germany** (0.565)	Czech R. (0.429)	France (0.611)	France (0.327)
4	France (0.183)	Denmark (0.428)	Belgium (0.597)	Italy (0.319)
5	Italy (0.100)	Bulgaria (0.428)	Italy (0.319)	**Germany** (0.302)
C	σ_{all}	$\sigma_{i/o}$	σ_{i+o}	σ_{comp_io}
1	Germany (1.000)	Brazil (1.000)	USA (1.000)	Brazil (1.000)
2	France (0.607)	Argentina (0.446)	UK (0.420)	Japan (0.773)
3	Canada (0.486)	Japan (0.303)	China (0.342)	**Russia** (0.398)
4	UK (0.339)	**Russia** (0.187)	Canada (0.317)	China (0.395)
5	**Russia** (0.265)	S. Korea (0.144)	India (0.297)	France (0.343)

The numbers in brackets represent the normalized strengths from Eqs. (1)–(4) on each country. Values in bold highlight countries which are relatively stable in ranking across all measurements

Finally, in the lower panel of Table 3, we obtain the greatest variation in rankings. The definition of σ_{i+o} produces the most intuitive results, in terms of overall industrial impact; $\sigma_{i/o}$ is again the least intuitive by placing less expected countries in the top 5. We consider σ_{all}, σ_{i+o} and σ_{comp_io} to be the most insightful in this third case.

5 Conclusions

In this paper we quantify the influence that builds between countries pertaining from particular groups, based on the amount of events occurring in one country, which are mentioned by other countries. The rationale is that if a national event is mentioned by a foreign country, then there is a quantifiable influence on that foreign country. We extract information about news mentions from around the world, using the Event Registry platform [8], corroborated with Gephi [2] and network science. The results are put together, forming directed weighted graphs of the so-called media influence.

We also introduce four different measures of node strength, based on well-known centrality metrics in a graph, and based on the different combinations of in- and out-degree. A close inspection of centrality metrics reveals that global-scale trade is decentralized, while smaller economical-systems, like the EU, are much more centralized by a few dominant countries. The differences in measured average degree and density make us conclude that inter-continental economics is more meritocratic in its structure, while intra-continental economical structures and mechanisms are more topocratic, namely advantaging the position and the in-betweenness of some countries.

For the best socio-political interpretation, it would be advisable to study multiple node centralities and have more appropriate background knowledge. Nevertheless, we hope to spark an interest in further interdisciplinary studies, like socio-politics, MOOCs and online education, which are based on the huge potential of Event Registry, and the prowess of social media analysis.

Acknowledgements Project supported by the Romanian National Authority for Scientific Research and Innovation (UEFISCDI), PN-II-RU-TE-2014-4-2040: "NOVAMOOC- Development and innovative implementation of MOOCs in Higher Education", and by the University Politehnica Timisoara "PCD-TC-2017" research grant.

References

1. Alberich, R., Miro-Julia, J., Rossl, F.: Marvel universe looks almost like a real social network. arXiv preprint cond-mat/0202174 (2002)
2. Bastian, M., Heymann, S., M., Jacomy, M.: Gephi: an open source software for exploring and manipulating networks. In: Proceedings of the 3rd Int'l Conference On Web And Social Media, vol. 8, pp. 361–362 (2009)
3. Bastian, M., Jacomy, M., Venturini, T., Heymann, S.: Forceatlas 2, a continuous graph layout algorithm for network visualization designed for the gephi software. PLoS One 9(6), e98679 (2014)
4. Blondel, V.D., l. Guillaume, J., Lambiotte, R., Lefebvre, E.: Fast unfolding of communities in large networks. J. Stat. Mech. **2008**, 10 (2008)
5. Copeland, B.R.: Strategic interaction among nations: negotiable and non-negotiable trade barriers. Can. J. Econ. **23**, 84–108 (1990)
6. de Zúñiga, H.G., Jung, N., Valenzuela, S.: Social media use for news and individuals' social capital, civic engagement and political participation. J. Comput. Mediat. Commun. **17**(3), 319–336 (2012)
7. Kugler, J., Domke, W.: Comparing the strength of nations. Comp. Pol. Stud. **19**(1), 39–69 (1986)
8. Leban, G., Fortuna, B., Brank, J., Grobelnik, M.: Event registry: learning about world events from news. In: Proceedings of the 23rd International Conference on World Wide Web, pp. 107–110. ACM, New York (2014)

9. Newman, M.E.: Modularity and community structure in networks. Proc. Natl. Acad. Sci. **103**(23), 8577–8582 (2006)
10. Newman, M.: Networks: An Introduction, pp. 1–2. Oxford University Press Inc., New York (2010)
11. Ricci, F., Rokach, L., Shapira, B.: Recommender systems: introduction and challenges. In: Recommender Systems Handbook, pp. 1–34. Springer, Berlin (2015)
12. Song, C., Havlin, S., Makse, H.A.: Self-similarity of complex networks. Nature **433**(7024), 392–395 (2005)
13. Teng, C.Y., Lin, Y.R., Adamic, L.A.: Recipe recommendation using ingredient networks. In: Proceedings of the 3rd Web Science Conference, pp. 298–307 (2012)
14. Topirceanu, A., Udrescu, M.: Fmnet: Physical trait patterns in the fashion world. In: Proceedings of the 2nd European Network Intelligence Conference, pp. 25–32. IEEE, New York (2015)
15. Topirceanu, A., Udrescu, M.: Statistical fidelity: a tool to quantify the similarity between multivariable entities with application in complex networks. Int. J. Comput. Math. **94**(9), 1787–1805 (2017)
16. Topirceanu, A., Udrescu, M., Vladutiu, M.: Network fidelity: a metric to quantify the similarity and realism of complex networks. In: Proceedings of the 3rd International Conference on Cloud and Green Computing, pp. 289–296. IEEE, New York (2013)
17. Topirceanu, A., Udrescu, M., Vladutiu, M.: Genetically optimized realistic social network topology inspired by facebook. In: Online Social Media Analysis and Visualisation, pp. 163–179. Springer, Cham (2014)
18. Topirceanu, A., Barina, G., Udrescu, M.: Musenet: collaboration in the music artists industry. In: Proceedings of the 1st European Network Intelligence Conference, pp. 89–94 (2014)
19. Topirceanu, A., Duma, A., Udrescu, M.: Uncovering the fingerprint of online social networks using a network motif based approach. Comput. Commun. **73**, 167–175 (2016)
20. Wang, X.F., Chen, G.: Complex networks: small-world, scale-free and beyond. IEEE Circuits Syst. Mag. **3**(1), 6–20 (2003)

Identifying Promising Research Topics in Computer Science

Rajmund Klemiński and Przemyslaw Kazienko

Abstract In this paper, we investigate an interpretable definition of promising research topics, complemented with a predictive model. Two methods of topic identification were employed: bag of words and the LDA model, with reflection on their applicability and usefulness in the task of retrieving topics on a set of publication titles. Next, different criteria for promising topic were analyzed with respect to their usefulness and shortcomings. For verification purposes, the DBLP data set, an online open reference of computer science publications, is used. The presented results reveal potential of the proposed method for identification of promising research topics.

Keywords Research prediction · Promising topic · Topic modelling · DBLP

1 Introduction

The study of science itself, its trends and underlying phenomena is a complex topic. Researchers have engaged in efforts to understand the evolution of scientific research areas, emergence of topics, and the structure of scientific research in both newly emerging and well-established areas. Perhaps the most exciting is the prospect of being able to accurately predict the rise and fall of individual research topics and areas. Such a look into the future is of great value to science and industry alike, plotting the path for future research activities. There have been many attempts at unveiling these future trends, rising in complexity of the solutions over the years. We attempt to take a step back and provide a process and results that are easy to understand, interpret, and take advantage of.

R. Klemiński (✉) · P. Kazienko
Department of Computational Intelligence, ENGINE - The European Centre for Data Science,
Faculty of Computer Science and Management, Wroclaw University of Science and Technology,
Wroclaw, Poland
e-mail: rajmund.kleminski@pwr.edu.pl; kazienko@pwr.edu.pl

© Springer International Publishing AG, part of Springer Nature 2018
R. Alhajj et al. (eds.), *Network Intelligence Meets User Centered Social Media
Networks*, Lecture Notes in Social Networks,
https://doi.org/10.1007/978-3-319-90312-5_16

2 Related Work

Prediction and identification of new, emerging or otherwise important topics has been researched for a while now. Interestingly, while "hot topic" as a term appears in many scientific articles, particularly the titles, it is more often than not self-proclaimed by the authors. Quantifiable research in this area, on the other hand, commonly focuses on online communities. This includes both detection [7] and propagation models [11]. There is a variety of approaches in the field of emerging topic detection. Scientists have used core documents [2], bibliometric characteristics [8], and Wikipedia sources [10] to identify and label new, emerging topics of research in different areas of science. Recently, complex approaches for predicting the trends and changes of the trends for topics have been developed with the use of ensemble forecasting [5] and rhetorical framing [9]. Evolution of the topics themselves has been tracked through metrics such as citations [3].

3 Data Sources

There are a number of possible sources of bibliographic data available for research, varying in scientific areas covered, the selection of sources indexed as well as the terms and conditions applying to accessing the full database. Considering that DBLP is the only publicly available data set from among the sources relevant for us (Table 1), we have decided to use it in our research. It was our reasoning that, as the database covers the areas of science related to computer science, it would allow for reviewing the results personally, without the need to engage outside-field experts. The DBLP database contains a variety of bibliographic information on publications, including the list of authors, a title, publication year, and journal/conference of publication. Additional details are available for select positions. Citation information, however, is missing.

4 Promising Topic

The notion of a promising research topic is well-known to every scientist; during our work, we encounter problems that appear "promising" to us. Such a feeling is usually driven by the experience in the field or based on the intuitive understanding of the problem in question. This makes the notion of a promising topic hard to translate into objective measures without which, in turn, it is impossible to conduct research on the problem of predicting promising topics. We have attempted to define a promising topic in two ways, using only the basic metrics available. While we recognize that such a nebulous term has more nuance, the research presented in this paper was meant to investigate the effectiveness of a simple approach, providing a reference point for the future work.

Table 1 Available bibliographic data sources

Name	Areas	Availability	Coverage	Temporal coverage
Scopus	All science	Licensed	Best journals and conferences	1995–present
Web of Science	All science	Licensed	Best journals and conferences	1900–present
PubMed	Biology and Medicine	Licensed	Best journals and conferences	1966–present
APS	Physics	By request	Best journals and conferences	1893–present
Google Scholar	All science	Scrapping	All web sources	Varies by field
DBLP	Computer science	**Public**	All web sources	1936–present

1. **Significance** A research topic is *significant* if there are many articles being written on it in a year, or a sizable community of scientists are involved in writing such papers. In this view, the more promising a topic, the higher value of the metric capturing these features in the following period.
2. **Growth** While the significance of a topic gives us knowledge of how widespread it is, there is an argument that can be made against it. Once a topic becomes the central focus of a sizable community of scientists, it usually retains that status for years to come. In that sense, the notion of significance is a stagnant one, favoring established areas of research over those that experience their rapid development. A metric focusing on the *percentage growth* of either the number of articles published or the quantity of community involved in the research should capture the nuances more accurately.

5 Topics

1. **Topic identification** The first step on the road to prediction of promising topics is to identify research topics present within the database of articles. Following that, each article in the database can be described by a subset of identified topics, denoting the research areas said paper pertains to. Similarly to our approach in the case of defining a promising topic, we have selected two methods of identifying topics, differing in complexity.
2. **Bag of words** The initial approach was to use a BoW model on the corpus of article titles obtained from the DBLP database. Considering the low value of uni-grams in the task of identifying research topics, we elected to only consider bi-grams. This is further rationalized by the fact that typically, a topic can be captured in a one- or two-word description.

3. **Latent Dirichlet allocation** LDA, a widely used statistical model designed to identify topics within the corpus, was the second method we have employed [1]. It is a significantly more complex solution compared to bag of words, introducing its own advantages and problems. In particular, as shown in [4], the LDA model suffers from a sparsity problem when used on a corpus of short texts. To combat this problem, we have modified our corpus according to our idea of n-gram corpus, further described in 3.
4. **Topic features** In order to attempt our predictions, we have calculated several features describing the topics identified within our corpus. Note that these are based on the information available in the DBLP data set which is relatively sparse when it comes to quantifiable data.

1. **Document frequency for topic (DFT)** Denotes the number of documents within the data set that were identified as a part of a given topic. This feature is calculated separately for every year considered in our research.
2. **ΔDFT** A relative change between the value of DFT between 2 years, calculated for each pair of consecutive years.
3. **Distinct author frequency (DAF)** Denotes the number of unique authors present in the set of all documents assigned to a given topic. This feature is calculated separately for every year.
4. **ΔDAF** Analogously to the ΔDFT, this feature expresses the relative change in DAF. Calculated for every pair of consecutive years.
5. **Term frequency-inverse document frequency (TF-IDF)** We have employed the popular in bag of words measure of TF-IDF for both of the models; it is easily achievable for the topics identified by the LDA model by a simple analogy. This feature is calculated separately for every year.
6. **Popularity** A feature describing how many documents pertained to the given topic, relative to all documents published during a given year. Calculated for every year separately.

6 Research Topic Prediction

Figure 1 showcases the prediction method applied in our research. Each step is further described in this section.

6.1 Corpus Processing

1. **Corpus preparation**. We begin the preparation of data by creating a corpus of document titles published between the specified years Y_1 and Y_n. This will serve as the basis for further processing.

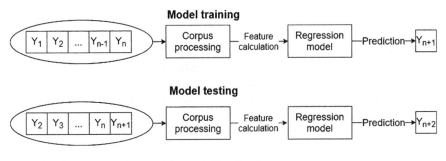

Fig. 1 Predictive process for promising topic identification

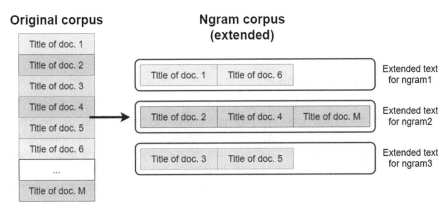

Fig. 2 Construction of the n-gram corpus

2. **Lemmatization**. In order to generalize article titles, we have decided to use a WordNet-based lemmatizer available for Python in [6]. Lemmatization was followed by filtering out of stop-words.

3. **N-gram corpus**. As mentioned previously, LDA model suffers from text sparsity if used on a collection of short texts. To combat this issue, we have constructed a special N-gram driven corpus. The idea behind this operation is simple. First, we acquire bi-grams from the document title corpus as a part of employing the bag of words model. Then, for each of the bi-grams found, we concatenate all document titles containing this bi-gram into a single position within the newly created n-gram corpus. This idea is visualized in Fig. 2. We have noticed a substantial improvement in both interpretability and cohesiveness of topics identified by LDA ran on n-gram corpus.

4. **Topic assignment**. Once the topic model has been trained on the corpus, we assign every probable topic to every document in the data set, allowing for later calculations. Note that this step is only carried out for LDA model.

6.2 Prediction

1. **Significance** Promising topic prediction in the sense of significance is achieved through the use of a regression model to predict either the DFT or DAF feature of topics, then rank them based on these features. To verify prediction accuracy, we compare a set of most promising topics as returned by our predictive model to real-world most promising topics.
2. **Growth** Promising topic prediction in the sense of growth is achieved through the use of a regression model to predict either ΔDFT or ΔDAF feature of topics. Further steps are the same as described above.

7 Experiments

7.1 Experimental Setup

1. **Scope** For our experiments, we have investigated three periods, starting with the value of Y_1 as 2007, 2008, and 2009. In each of these cases, n was equal to 5. What it does mean for the model is, that we train the regression model based on years 2007–2011 predicting year 2012, then test the model by using years 2008–2012 to predict year 2013. We perform analogous task for the remaining two values of Y_1.
2. **Regression model and features** For the purposes of this experiment, we have chosen a linear regression model with feature selection. This choice was motivated by the ease of interpretation provided. There was a total of 28 input features prior to the feature selection process: DFT, DAF, $TF\text{-}IDF$, and $Popularity$ present five times each, once for every year, while ΔDFT and ΔDAF were present four times each calculated for every consecutive pair of years in the input data set.
3. **Verification** To verify the accuracy of our predictions, we rank the set of articles published in year Y_{n+2} based on the predicted value of a desired feature (for example, DFT) and the real value of the same feature. We then select top x positions from both rankings and acquire the common part of these two sets. Note that such operation disregards the precise order of the ranking. Our rationale is that it is more valuable to accurately predict that a specific topic will be among the top 20 in the following year than whether it will be 5th or 10th. The final accuracy score is the proportion of the number of topics predicted to be in the top x compared to real top x topics.

7.2 Results

1. **Bag of words** The bag of words model achieved a reasonable accuracy when predicting promising articles by the value of their DFT feature, as shown in Fig. 3a. Note that the accuracy is much lower for year 2015 than the two preceding years. This might be related to the fact that DBLP database did not include arXiv articles prior to 2015. A sudden change in the volume of documents and, presumably, the distribution of words in their titles would be expected to lower the predictive ability of the model. Interestingly, prediction focused on the size of a community (DAF, Fig. 3b), remained unaffected. The situation is different for the case of prediction based on the community growth (Fig. 3d) as it suffers from lesser consistency prior to the year of 2015. Growth measure in regards to the number of documents published on a given topic is predicted with lower variation of accuracy (Fig. 3c). In both cases the predictive model fares much worse than when operating on raw numbers.
2. **LDA** Predicting promising research topics in the sense of a significance yields very accurate results, as seen in Fig. 4a, b. This aligns with the results achieved through the use of a bag of words model. In the case of both DFT and DAF feature, the linear regression model is capable of good predictions that remain consistent within the scope of a year. We do not observe either of the features to be predicted with visibly lower accuracy, unlike in the previous case. This time, however, the year of 2014 scores the lowest across the lengths of rankings and features as well. This accuracy drop is not as dramatic as in the case of bag of words model, but consistent enough to be visible. The growth metrics are predicted with a significantly lower accuracy, evidenced in Fig. 4c, d. Attempts of predicting any of the two features considered as indicative of growth fail to produce consistent results; for each of the three publishing years taken into consideration, it is either the prediction of a ΔDFT or ΔDAF feature that achieves varied accuracy between the length of a ranking considered. In some cases the most discriminating subset, top ten research topics, scores 0% accuracy. This is considerably worse than the results of the same task carried out with the use of a bag of words model.

7.3 Discussion

The stark difference between results of our predictions for raw numbers and the growth of such numbers can be explained intuitively. As mentioned before, once a topic achieves a wide following, it is likely to remain a sizable research area for a long time. This introduces little dynamic to the top rankings, as they can be expected to bear similarity to each other between the years. Changes in numbers, on the other hand, are relative and capture more information. Thus, it would be expected that a task of such prediction would be harder.

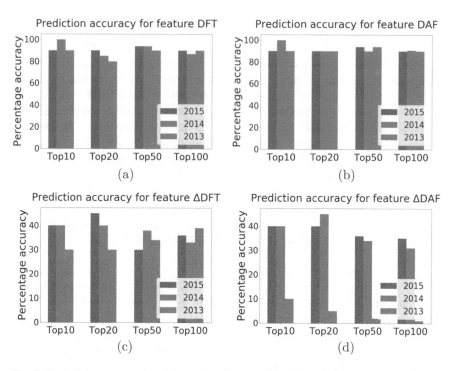

Fig. 3 Prediction accuracy using the bag of words model. (**a**) DFT prediction. (**b**) DAF prediction. (**c**) ΔDFT prediction. (**d**) ΔDAF prediction

The metrics available within the database are crucial. When stripped of purely bibliographical information, the DBLP data set contains little quantifiable information; features proposed and included in our research were computed based on the records in the data set. As shown by the predictive results presented, operating on basic features might yield results when attempting to predict other simple features, but is insufficient to capture true nuance. In such a case, the benefit of high interpretability is of comparably little value.

It is impossible to discuss the results of our research without retrieving the most influential bi-grams and research topics from our topic identification models. Accuracy scores can be high, but a manual review of the supposed most promising research areas is necessary. For this reason, we present two most promising topics retrieved by LDA model, with regard to ΔDFT in Fig. 5 and a list of five most promising bi-grams, by the same metric, in Table 2.

Topics retrieved by the LDA model appear anomalous; while there seems to be an internal connection between the words, these are either not hinting towards a realistic area of research or are mixed with one or more similarly related groups. A notable problem is the abundance of words common in computer science, such as "system" or "application." These would belong to a list of stop-words in an ordinary case. However, through the course of several experiments we have determined that

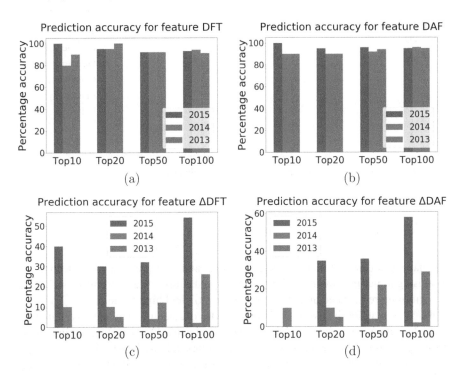

Fig. 4 Prediction accuracy using the LDA model. (**a**) DFT prediction. (**b**) DAF prediction. (**c**) ΔDFT prediction. (**d**) ΔDAF prediction

Fig. 5 Word clouds of top two topics retrieved by the LDA model

no matter the extent of a stop-list, there would be a stable number of about 20 words assigned to anywhere between 5 and several hundred topics. This leads us to the conclusion that it is extremely hard to retrieve topics from titles of scientific articles alone. This seems in line with what we can observe in scientific literature, as the titles shifted from being purely informative towards attracting the potential reader with a unique description of the problem.

While bi-grams are more suited for identifying research areas from titles, they also prove to have limitations. Topics like "deep learning" can be artificially over-represented, as they commonly attract additional attention to the work. Unlike more elaborate mechanisms, using a BoW approach doesn't allow for identifying a topic

Table 2 Top 5 most
promising bi-grams with
respect to ΔDFT

Rank	Topic
1	Deep learning
2	Data analytics
3	Deep neural
4	Massive system
5	Full duplex

when nuanced or metaphorical wording is used during the title-creation, leading to inserting one or more words into an established phrase.

8 Conclusions

We have presented our approach for simple and interpretable identification of promising topics in computer science. The approach itself shows promise, but seems to require more nuanced and in-depth features to yield high-value results. We have highlighted that, for the use of LDA model, titles of scientific publications cannot be treated as any other short text. We believe this is caused by scientific community's growing awareness of the benefits that higher marketability brings in.

In the future work, we will aim at acquiring a data set containing advanced information to measure its impact on our predictive abilities. Furthermore, we are interested in measuring in higher detail the impact a community working on a scientific topic can have on the topic itself.

Acknowledgements This work was partially supported by the National Science Centre, Poland, project no. 2016/21/B/ST6/01463 and by European Union's Horizon 2020 research and innovation programme under the Marie Skłodowska-Curie grant no. 691152 (RENOIR); by the Polish Ministry of Science and Higher Education fund for supporting internationally co-financed projects in 2016-2019, no. 3628/H2020/2016/2.

References

1. Blei, D.M., Ng, A.Y., Jordan, M.I.: Latent dirichlet allocation. Adv. Neural Inf. Proces. Syst. **1**, 601–608 (2002)
2. Glänzel, W., Thijs, B.: Using 'core documents' for detecting and labelling new emerging topics. Scientometrics **91**(2), 399–416 (2012)
3. He, Q., Chen, B., Pei, J., Qiu, B., Mitra, P., Giles, L.: Detecting topic evolution in scientific literature: how can citations help? In: Proceedings of the 18th ACM Conference on Information and Knowledge Management, CIKM '09, pp. 957–966 (2009)
4. Hong, L., Davison, B.D.: Empirical study of topic modeling in twitter. In: Proceedings of the First Workshop on Social Media Analytics, SOMA '10, pp. 80–88 (2010)
5. Hurtado, J.L., Agarwal, A., Zhu, X.: Topic discovery and future trend forecasting for texts. J. Big Data **3**(1), 7 (2016)

6. Loper, E., Bird, S.: Nltk: The natural language toolkit. In: Proceedings of the ACL-02 Workshop on Effective Tools and Methodologies for Teaching Natural Language Processing and Computational Linguistics - Volume 1, ETMTNLP '02, pp. 63–70 (2002)
7. Lu, Y., Zhang, P., Liu, J., Li, J., Deng, S.: Health-related hot topic detection in online communities using text clustering. PLoS One **8**(02), e56221 (2013)
8. Mund, C., Neuhäusler, P.: Towards an early-stage identification of emerging topics in science–the usability of bibliometric characteristics. J. Informetrics **9**(4), 1018–1033 (2015)
9. Prabhakaran, V., Hamilton, W.L., McFarland, D.A., Jurafsky, D.: Predicting the rise and fall of scientific topics from trends in their rhetorical framing. In: ACL (2016)
10. Wang, Y., Joo, S., Lu, K.: Exploring topics in the field of data science by analyzing Wikipedia documents: a preliminary result. Proc. Am. Soc. Inform. Sci. Technol. **51**(1), 1–4 (2014)
11. Zhang, B., Guan, X., Khan, M.J., Zhou, Y.: A time-varying propagation model of hot topic on {BBS} sites and blog networks. Inform. Sci. **187**, 15–32 (2012)

Index

A
Acceleration coefficient (ACC), 87–90
Arules algorithm, 164
Assortativity coefficient (AC), 123
Assumptions of PDBC
    central nodes, 31–32
    equal amount of flow, 25–26
    measure correlations, 29–31
    process moves between every node pair, 24–25
    process moves on shortest paths, 24
Average path length (APL), 123

B
Bag of words (BoW) model, 233, 237
Barabasi-Albert model, 122, 126
Behavior-based relevance estimation model (BBRE), 38
    active edge, 41
    behavioral changes, 36
    distribution patterns, 44
    existing interaction, 36
    experiments, 41–44
    exponential decay, 41
    geoMeanOfChangeInMean, 40
    geometric mean, 39
    inactive edge, 41
    link prediction, 35, 36
    recency of interactions, 35
    rolling mean, 39
    sliding windowed mean, 38
    social decay, 36, 40

    stable interactions, 43
    Western Electric Rules, 37
Bootstrap percolation, hyperbolic networks
    activation mechanism, 10
    description, 4–5
    discrete-time process, 5
    graphical creation, 9–10
    heatmap, 12
    NP-hard, 5
    proposed method, 8
    random recovery, 11, 13
    SI model, 5
    SIS model, 6
    targeted recovery, 11, 13–14

C
Change detection ratio (cdr), 139
Claudio Rocchini, 6–7
Clustering coefficient (CC), 123
Community detection
    clustering average, 150, 151
    complexity, 149–150
    density based optimization, 147–148
    directed graph, 144
    Louvain algorithm, 150, 151
    Markov chain, 148–149
    NMI average, 150
    NMI scores, 146
    pattern based distance, 149
    selection of representatives, 148
    select partition, 149
    Sponge walker, 144

© Springer International Publishing AG, part of Springer Nature 2018
R. Alhajj et al. (eds.), *Network Intelligence Meets User Centered Social Media Networks*, Lecture Notes in Social Networks,
https://doi.org/10.1007/978-3-319-90312-5

Printed in the United States
By Bookmasters